ADVANCES IN HEAT TRANSFER

Volume 41

Advances in
HEAT TRANSFER

Serial Editors

George A. Greene
Energy Sciences and Technology
Brookhaven National Laboratory
Upton, New York

Young I. Cho
Department of Mechanical Engineering and Mechanics
Drexel University
Philadelphia, Pennsylvania

Avram Bar-Cohen
Department of Mechanical Engineering
University of Maryland
College Park, Maryland

Volume 41

Founding Editors

Thomas F. Irvine, Jr. *State University of New York at Stony Brook, Stony Brook, NY*
James P. Hartnett *University of Illinois at Chicago, Chicago, IL*

AMSTERDAM • BOSTON • HEIDELBERG • LONDON
NEW YORK • OXFORD • PARIS • SAN DIEGO
SAN FRANCISCO • SINGAPORE • SYDNEY • TOKYO
Academic Press is an imprint of Elsevier

Academic Press is an imprint of Elsevier
525 B Street, Suite 1900, San Diego, CA 92101-4495, USA
30 Corporate Drive, Suite 400, Burlington, MA 01803, USA
32 Jamestown Road, London NW1 7BY, UK
Radarweg 29, PO Box 211, 1000 AE Amsterdam, The Netherlands

First edition 2009

Copyright © 2009 Elsevier Inc. All rights reserved

No part of this publication may be reproduced, stored in a retrieval system
or transmitted in any form or by any means electronic, mechanical, photocopying,
recording or otherwise without the prior written permission of the publisher

Permissions may be sought directly from Elsevier's Science & Technology Rights
Department in Oxford, UK: phone (+44) (0) 1865 843830; fax (+44) (0) 1865 853333;
email: permissions@elsevier.com. Alternatively you can submit your request online
by visiting the Elsevier web site at http://www.elsevier.com/locate/permissions, and selecting
Obtaining permission to use Elsevier material

Notice
No responsibility is assumed by the publisher for any injury and/or damage to persons
or property as a matter of products liability, negligence or otherwise, or from any use
or operation of any methods, products, instructions or ideas contained in the material
herein. Because of rapid advances in the medical sciences, in particular, independent
verification of diagnoses and drug dosages should be made

British Library Cataloguing in Publication Data
A catalogue record for this book is available from the British Library

Library of Congress Cataloging-in-Publication Data
A catalog record for this book is available from the Library of Congress

ISBN: 978-0-12-381424-1
ISSN: 0065-2717

For information on all Academic Press publications
visit our website at elsevierdirect.com

Printed and bound in USA
09 10 11 12   10 9 8 7 6 5 4 3 2 1

Working together to grow
libraries in developing countries

www.elsevier.com | www.bookaid.org | www.sabre.org

ELSEVIER   BOOK AID International   Sabre Foundation

CONTENTS

Contributors vii

Preface ix

Microgrooved Heat Pipe

B. Suman

I. Introduction	1
A. Brief Overview of Chapter	1
B. Brief History of Heat Pipes	1
C. Aim and Scope	3
D. Heat Pipe	3
E. Microgrooved Heat Pipe	5
II. Applications of a Micro Heat Pipe	7
A. Electronic Cooling	7
B. Human Disease Remedy	8
C. Spacecraft Thermal Control	9
D. Solar Energy Conversion	10
E. Aircraft Temperature Control	10
F. Fuel Cell Temperature Regulation	10
III. Steady-State Operation of Micro Heat Pipes	11
A. Literature Review of Various Steady-State Models	11
B. Steady-State Experimental Studies	27
C. Electrohydrodynamically Augmented Micro Heat Pipe	31
D. Steady-State Sensitivity Analysis	32
IV. Transient Operation of Micro Heat Pipes	33
A. Transient Models	33
B. Transient Experimental Study	41
C. Transient Sensitivity Analysis	43
V. Operating and Design Parameters	44
A. Fill Charge	44
B. Liquid Height	47
C. Coolant Liquid Temperature	47
D. Effect of Surface Tension Gradient on a Micro Heat Pipe	47
E. Friction Factor and Surface Roughness	48
F. Fabrication of Micro Heat Pipes	49
G. Performance Factor	52
VI. Design of Microgrooved Heat Pipes	53
A. Limitations to Heat Transport	54

VII. Future Direction.................................... 64
VIII. Conclusions 69
Nomenclature.. 71
References 72

A Review of Heat Transfer in Nanofluids

SARIT K. DAS and STEPHEN U.S. CHOI

I. Introduction.. 81
II. From Slurries to Nanofluid.......................... 82
III. Nanofluids: Novel Features 85
IV. Synthesis of Nanoparticles 87
V. Particle Characterization............................ 90
VI. Preparation of Nanofluids........................... 92
    A. Physical Dispersion Technique 93
    B. Chemical Dispersion Method 93
    C. Single-Step Methods............................ 94
VII. Thermal Conductivity Enhancement in Nanofluids 95
    A. Ceramic Nanofluids 95
    B. Metallic Nanofluids 100
    C. Carbon and Polymer Nanotube Nanofluids........ 104
VIII. Temperature Effect................................. 107
IX. Theories on Thermal Conductivity of Nanofluids 114
X. Convection in Nanofluids............................ 126
    A. Forced Convection in Nanofluids 126
    B. Experimental Works on Convection in Nanofluids.... 128
    C. Mechanisms in Convection of Suspensions 136
    D. Analytical and Numerical Studies on Convection in NanoFluids 140
    E. Natural Convection in Nanofluids................ 165
XI. Boiling in Nanofluids............................... 175
    A. Pool Boiling Heat Transfer at Higher Solid Particle Concentrations..... 176
    B. Pool Boiling Heat Transfer at Lower Solid Particle Concentrations 182
XII. Studies on CHF in Pool Boiling 184
XIII. Applications of Nanofluids.......................... 185
XIV. Summary and Future Direction of Research 187
Nomenclature.. 191
References 191

Author Index 199

Subject Index 203

CONTRIBUTORS

*Numbers in parentheses indicate the pages on which
the author's contributions begin.*

STEPHEN U. S. CHOI (81) Department of Mechanical and Industrial Engineering (MC 251), Nanofluids Laboratory, University of Illinois at Chicago, Chicago, Illinois 60607, USA

SARIT K. DAS (81) Department of Mechanical Engineering, Heat Transfer and Thermal Power Laboratory, Indian Institute of Technology Madras, Chennai, Tamil Nadu 600 036, India

BALRAM SUMAN (1) Department of Chemical Engineering and Materials Science, 151 Amundson Hall, 421 Washington Avenue SE, University of Minnesota, Minneapolis, MN 55455, USA

PREFACE

For more than 40 years, *Advances in Heat Transfer* has filled the information gap between regularly published journals and university-level textbooks. The series presents review articles on topics of current interest, starting from widely understood principles and bringing the reader to the forefront of the topic being addressed. The favorable response by the international scientific and engineering community to the 41 volumes published to date is an indication of the success of our authors in fulfilling this purpose.

In recent years, the editors have published topical volumes dedicated to specific fields of endeavor. Examples of such volumes are Volume 22 (*Bioengineering Heat Transfer*), Volume 28 (*Transport Phenomena in Materials Processing*), Volume 29 (*Heat Transfer in Nuclear Reactor Safety*), and Volume 40 (*Transport Phenomena in Plasma*). The editors intend to continue publishing topical volumes as well as the traditional general volumes in the future. Volume 32, a cumulative author and subject index for the first 32 volumes, has become a valuable tool to search the series for contributions relevant to their current research interests.

The editorial board expresses its appreciation to the contributing authors of Volume 41, who have maintained the high standards associated with *Advances in Heat Transfer*. Finally, the editors would like to acknowledge the efforts of the staff at Academic Press and Elsevier, who have maintained the attractive presentation of the volumes over the years.

Microgrooved Heat Pipe

B. SUMAN

Department of Chemical Engineering and Materials Science, 151 Amundson Hall, 421 Washington Avenue SE, University of Minnesota, Minneapolis, MN 55455

I. Introduction

A. Brief Overview of Chapter

A microgrooved heat pipe (MHP) is a reliable and efficient heat transport device for thermal control of integrated electronic circuits packaging, laser diodes, photovoltaic cells, infrared (IR) detectors, and space vehicles. The device transfers heat using evaporation and condensation of the coolant liquid, and the circulation of the coolant fluid is due to the capillary pressure generated due to the liquid meniscus difference at the evaporator and condenser sections. In this chapter, we present a comprehensive review on the MHP starting from its introduction of its initial concept. The following section first describes the historical development of this system. Since its introduction, several mathematical models and experimental studies have been presented to understand the capillary behavior in the MHP. Thus, numerous analytical, numerical, and experimental studies are presented to determine the fundamental parameters that govern the operation of micro heat pipes (MHPs) with an aim to provide state-of-art knowledge to both individuals, new to the MHP research and to those already working in the area of MHP. The development of microgrooved fabrication and various fabrication methods on the substrate have also been presented. In addition, the critical heat inputs, dryout length, fill charge, liquid temperature, performance factor, various heat pipe limitations, and design have been discussed. Finally, the future research direction in the area of MHPs has been outlined. Thus, this chapter could serve as a basis for heat pipe design, understanding the MHP concept, improving the performance, and for further expansion in its applications.

B. Brief History of Heat Pipes

A heat pipe is a sealed vessel used for transferring heat from one place to another using interfacial free energy gradients to control fluid flows in the absence of mechanical pumps. The concept of a heat pipe originated more than one and a half centuries ago when King and Perkins registered their patents [1,2]. The device was popularly known as the Perkins tube, which utilized either single or two-phase processes to transfer heat from a furnace to a boiler. They then proposed a portable oven in 1867. Gay [3] obtained a patent on a device similar to the Perkins tube, in which a number of vertical tubes were arranged with an evaporator located below a condenser. These devices, which were actually classified as thermosiphon heat exchanger, laid the groundwork for the development of a heat pipe. Schmidt proposed the use of ammonia or carbon dioxide near its critical points for filling thermosiphon tubes. Later, Gaugler [4] proposed a two-phase closed thermosiphon tube, which was a heat transfer device incorporating a wick or porous matrix for a capillary action. The wick or the porous medium used to give the driving force for fluid flows. Trefethen [5] resurrected the idea of a heat pipe in connection with the space program in 1962. In 1963, Grover [6] filed a U.S. patent application on behalf of the U.S. Atomic Energy Commission for an evaporation–condensation device and used the term "Heat Pipe" for the first time. Grover *et al.* [7] built several prototype heat pipes, the first of which used water as a working fluid and was soon followed by a sodium heat pipe which operated at 1100 K. The heat pipe concept received relatively little attention until Grover *et al.* [7] published the results of an independent investigation. Since then, heat pipes were employed in various applications, industries and labs. By 1969, there was a vast amount of interest on the part of NASA, Hughes, the European Space Agency, and other aircraft companies for regulating temperature in aircraft by the use of heat pipes.

Chi [8], Dunn and Reay [9], Terpstra and Van Veen [10], Peterson [11], Faghri [12], and Peterson [13] published several mathematical models, theories, and applications of different heat pipes. The aim of such models and theories were to determine the fundamental parameters that govern heat pipe operations [14–34]. Peterson [35] presented a review on the MHP research starting from the introduction of its initial concept. In addition to the experimental results and procedures, a summary of the analytical and numerical techniques used to model and the performance of these integral MHPs was also presented. Itoh *et al.* [36] presented a review on current development and application of an MHP in 1993. Starting from Peterson's first review on an MHP [37] in 1992, he updated the literature by writing another review in 1996 [38]. This has then been followed by Suman [39]. Apart from a series of reviews in *Applied Mechanics Reviews* by Peterson and Suman, Cao and Faghri [40] have reviewed the literature on MHPs till 1994. Shyu Ruey-Jong [41] has performed a review of heat pipe work in Taiwan with little attention on an MHP. In 1998, Peterson *et al.* [42] presented a review article, apart from a series of reviews in *Applied Mechanics Reviews*, on an MHP for both individuals new to the field and to those already working in this area.

C. AIM AND SCOPE

Various reviews were published on the MHP research, but none of them covered starting from its initiation of concept to today in detail. However, the substantial research was done on the MHP and all are interconnected. Therefore, this chapter presents a comprehensive review on the MHP research and development with an aim to give state-of-the-art knowledge to individuals new to the MHP research as well as to those already working in this area. The main objective is to review the previous experiments and models that were used or could be used for the analysis of MHPs. An overview of theoretical and experimental results for the analysis of MHPs has been presented. The steady-state and transient models have been presented along with their sensitivity analyses. The experimental investigations of steady-state and transient operations and various fabrication methods for microgrooves on substrate have also been presented. Critical heat inputs, dry-out length, fill charge, temperature of the liquid, performance factor, various heat pipe limits, and design of MHP have been discussed. Finally, the direction of future research in the area of MHPs has been identified and delineated. We would first briefly discuss a heat pipe and then we continue with the MHP.

D. HEAT PIPE

A heat pipe is a sealed vessel, which is used for transferring heat from one place to another. The heat pipe is filled with a working fluid and the pipe is sealed. Heat flux is applied to a portion of a heat pipe, called the evaporative section, to vaporize the liquid in that region. The vapor is pushed toward a section where the heat is taken out and is called the condenser section. Between the evaporative and condensing sections, there is a heat transport section called the adiabatic section. The capillary pressure and the temperature difference between the evaporator and the condenser sections promote the flow of the working fluid from the condenser back to the evaporator through the corner regions. It dissipates energy from heat source by the latent heat of evaporation in a nearly isothermal operation. Circulation of a working fluid inside the heat pipe is accompanied by phase changes at both the evaporator and the condenser. Therefore, it is also called a two-phase convection device.

1. Working Principle of a Heat Pipe

Basically, a heat pipe operates on a closed two-phase cycle and utilizes the latent heat of vaporization to transfer heat with very small temperature gradients. A schematic of the working principle of a heat pipe is presented in

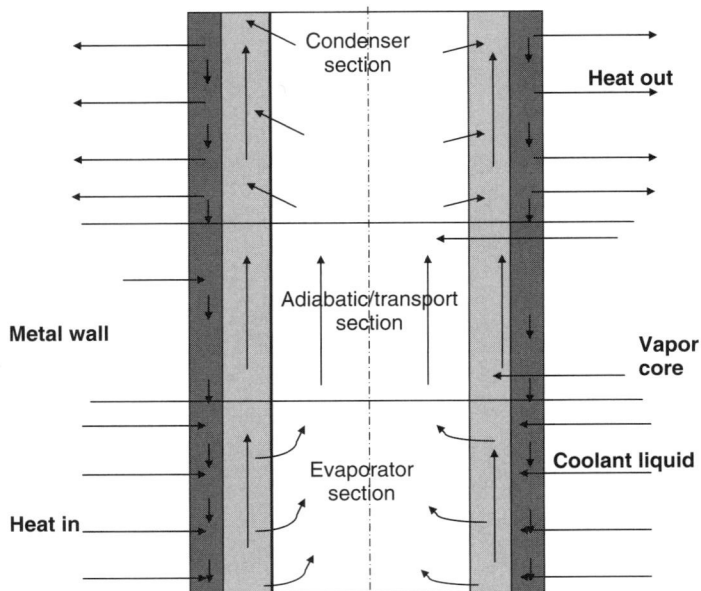

FIG. 1. Schematic of the working principle of a heat pipe.

Fig. 1. In a heat pipe, a working liquid is in equilibrium with its own vapor. The length of the heat pipe is divided into three parts: evaporative, adiabatic (transport), and condenser sections. Heat applied to the evaporator section by an external source is conducted through the pipe wall and the wick structure where it vaporizes the working fluid. The resulting vapor pressure drives the vapor from the adiabatic section to the condenser, where the vapor condenses, releasing its latent heat of vaporization to the provided heat sink. The capillary pressure generated by either the menisci in the wick or the difference in the liquid curvatures, pumps the condensed fluid back to the evaporator section. Therefore, the heat pipe can continuously transport latent heat of vaporization from the evaporator to the condenser section. As the latent heat of evaporation is usually very large, considerable quantities of heat can be transported with a very small temperature difference across the length of the heat pipe. The amount of heat that can be transported as latent heat of vaporization is usually a few orders of magnitude larger than the sensible heat in a conventional convective system with an equivalent temperature difference. Therefore, relatively large amounts of heat can be transported with small lightweight structures. The performance of a heat pipe is often expressed in terms of equivalent thermal conductivity. The large effective thermal conductivity of the heat pipes can be illustrated by the following examples. A heat pipe using water as the working fluid and operating at 150°C would have a thermal conductivity several hundred times than that of a copper bar of dimensions same as of the heat pipe [43–46].

2. Advantages of a Heat Pipe

The use of interfacial free energy gradients to control fluid flow naturally leads to a simpler and lighter heat transfer system, which is due to the absence of mechanical pumps. Therefore, passive engineering systems based on this principle are promising candidates for many applications such as the space program and electronic cooling. The heat transfer capacity of a heat pipe is several orders of magnitude greater than the best solid conductors because heat pipe operates on a closed two-phase cycle. A lithium-filled heat pipe developed at Los Alamos in mid-1980s transferred heat energy at a power density of 23 kW cm^{-2}. This results in a relatively small thermal resistance and allows physical separation of the evaporator and the condenser without having a high overall temperature drop. The promising feature of a heat pipe is the two-phase heat transfer mechanism and its working by the capillary phenomena with a lower pressure difference. The second feature makes it even more promising since there are devices which use two-phase heat transfer mechanism but they need a pressure difference of more than 5 atm. Another advantage of a heat pipe is its ability to increase the rate at which the working fluid is vaporized without a significant increase in the operating temperature when the heat flux in the evaporator is increased. Thus, heat pipes can function as a nearly isothermal device by adjusting the evaporation rate to accommodate a wide range of power inputs, while maintaining a relatively constant temperature. In addition, the heat pipes have their small thermal response time. The thermal response time is considerably less than the other kinds of heat transfer devices in general and solid conductors in particular since heat pipes utilize a closed two-phase cycle. This response time is not a function of length [47]. As a result of these advantages and because of its high reliability, stand-alone operation, and minimum maintenance, heat pipes have very wide applicability. A number of heat pipes have been developed for numerous cooling applications such as microelectronics chip, space, reactor, and engine.

E. MICROGROOVED HEAT PIPE

MHPs, as proposed by Cotter [48], are defined to be so small that the mean curvature of the liquid–vapor interface is necessarily comparable in magnitude to the reciprocal of the hydraulic radius of the total flow channel. An MHP should be distinguished from miniature heat pipes which have a larger hydraulic radius to liquid–vapor interface curvature ratio. A typical

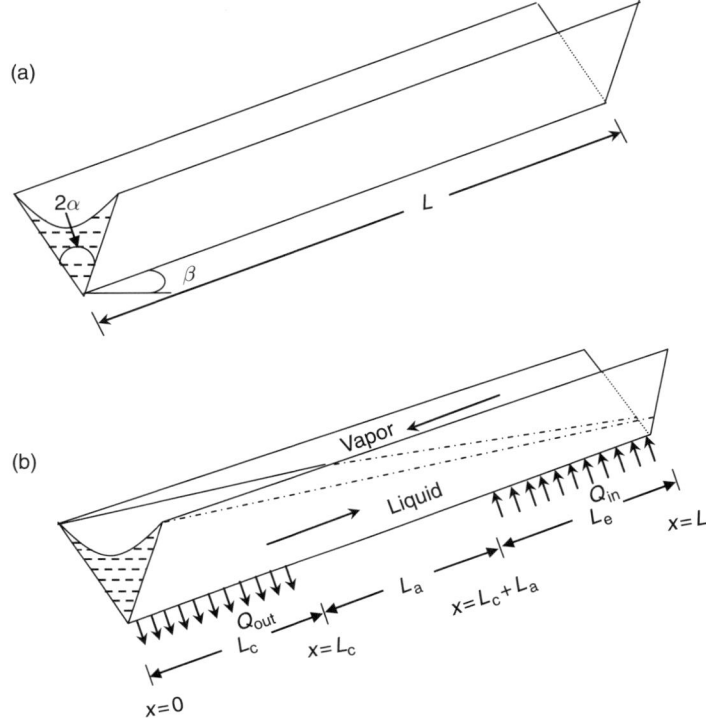

FIG. 2. (a) Schematic of a V-shaped MHP showing groove geometry. (b) Schematic of a V-shaped MHP showing flow directions.

MHP has a hydraulic diameter ranging from 10 μm to several millimeters and a length of up to several centimeters. MHP devices range in size from 1 mm in diameter and 60 mm in length to 30 mm in diameter and 10 mm in length. By definition, an MHP requires the Bond number $(B_o = \rho g R_h^2/\sigma)$ to be on the order of ≤1. A schematic of a V-shaped MHP has been presented in Fig. 2 (a) and (b) showing groove geometry and flow directions, respectively. The maximum heat flux dissipated with these devices is 60 W cm^{-2} [49].

1. Working Principle of a Micro Heat Pipe

Fundamental operating principles of an MHP are essentially the same as those occurring in the conventional heat pipes. The vaporization and condensation processes cause the liquid–vapor interface in the channels to change continuously along the heat pipe and results a curvature difference between the farthest ends of an evaporative (hot end) and a condensing (cold end) section in a V-grooved MHP. No wicking structure is required since the flow of liquid is due to the curvature difference between the hot and the cold ends. Therefore, it can be directly grooved on solid substrates. In the earth environment, these heat pipes usually have very small characteristic dimensions to keep the Bond number small and are known as MHPs. Outside the earth environment, in microgravity, the characteristic dimension for a small Bond number can be considerably large. These larger systems should accommodate larger heat fluxes due to smaller viscous losses. The MHP has become more attractive because of its high efficiency, reliability, and cost effectiveness. Some of the applications of an MHP have given in the next section.

II. Applications of a Micro Heat Pipe

A. Electronic Cooling

The work on heat transfer in microelectromechanical systems, which is the basement of a micro heat pipe, is inspired by Tuckerman and Pease's pioneering contribution [50]. As the electronic devices increase in computing ability and processing speed and decrease in size, there is a problem of efficient removal of the heat generated. Excess heat produces stress on the internal components of the computer, and thus creates reliability problems. Fan-assisted heat sinks require electrical power and they reduce life of battery. Standard metallic heat sinks, capable of dissipating the heat load, are too large to be incorporated into the notebook packages. Two-phase heat transfer mechanisms are found to be a viable option for such purposes. Thus, MHPs offer a highly efficient passive, compact heat transfer solution. The 3- to 4-mm-diameter heat pipes can effectively remove the high flux heat from the processor. The heat pipe spreads the heat load over a relatively large area heat sink, where the heat flux is so low that it can be effectively dissipated through the notebook case to ambient air. The heat sink can be the existing components of the notebook, from electromagnetic interference shielding under the key pad to metal structural components. Hence, currently, one of the best applications for heat pipes is cooling the Pentium processors, in notebook computers [51]. Due to the limited space and power available in notebook computers, heat pipes are ideally suited for cooling the high-power chips [52–56].

Both micro and miniature heat pipes appear to be promising for the use in microelectronics cooling. MHPs, which could be embedded directly onto the silicon substrate of an integrated circuit, have been investigated in several studies conducted at Texas A&M (Mallik et al. [14,15]). Babin et al. [15], also at Texas A&M, reported experimental data taken on several water-charged

MHPs with a cross-sectional dimension of about 1 mm. Adkins *et al.* [57] at Sandia National Laboratories-Albuquerque discussed the use of a "heat-pipe heat spreader" embedded in a silicon substrate as an alternative to the conductive cooling of integrated circuits using diamond films. A technique has been developed by which very small MHPs, approximately 35 μm in diameter, can be fabricated as an integral part of semiconductor devices. These heat pipes function as highly efficient heat spreaders collecting heat from the localized hot spots and dissipating the heat over the entire chip surface. Incorporation of these heat pipes as an integral part of silicon wafers has been shown to significantly reduce the maximum wafer temperature and reduce the temperature gradients occurring across these devices [58]. To cite another application of micro and miniature heat pipes, a novel bellows heat pipe to conduct heat away from electronic components to a heat sink has been developed. The heat pipe operates as a thermal switch to cool electronic components and insures that the temperatures in the components do not fall below or exceed predetermined levels. This design can be used as a stand-alone mechanism for individual microelectronic components in multicomponent modules or on a larger scale as a flexible thermal contact for electronic equipment or systems [59]. Kojima *et al.* [59] used high efficient cooling methods like a H_2O cooling and a blower forced cooling unit is demanded on transmission systems, because of high large-scale integrated circuits (LSIs) used. Heat pipes play a crucial role on such a condition. They developed new cooling systems with MHPs and studied its thermal characteristics. On a printed circuit board, the assembly of cooling system was mounted on every surface of LSIs and cooling fins were connected with MHPs. The ability of cooling power was three or five times as high as the conventional one.

A new technique has been developed to overcome the capillary limitation normally encountered in extremely long heat pipes. This technique when combined with flexible heat pipes having variable cross-sectional area can provide a highly efficient method for removing and dissipating the heat generated in high-power electronic packages developed for the computer industry [60]. Both stock and custom heat pipes for a variety of electronic cooling applications is manufactured in California. Des Champs Laboratories, Incorporated, of Natural Bridge Station, Virginia, manufactures standard and custom industrial air-to-air heat pipe heat exchangers for electronic cabinet cooling. North [61] and Rosenfeld [62], both at Thermacore, discuss some applications of heat pipes to microelectronics cooling [63].

B. HUMAN DISEASE REMEDY

These days, the applications of a micro-grooved heat pipe have been extended from the electronic cooling and space program to a remedy of many diseases. Peterson and his coworkers are the world leader in this direction. In this section, we will briefly discuss his contributions on the use of an MHP in biological applications.

Fletcher and Peterson [64] suggested that an MHP catheter of the size of a hypodermic needle (~1 mm) may be charged with an appropriate working fluid to assure a constant temperature operation within a therapeutic temperature range of 42.5°C (108.5°F) to 43°C (109.4°F) for the use in hyperthermia cancer treatments. The MHP has a promise to provide a controllable heat transfer rate at a constant temperature and may be matched to the thermal conductivity of the tissue and the degree to which the cancerous tumor is perfuse. In this way, the MHP catheter [64] may be used to treat cancerous tumors in body regions which can not be treated by other means. This MHP catheter may also be used for cryosurgery applications with appropriate support facilities.

Peterson and Fletcher [65] developed and designed temperature control mechanisms for the use in conjunction with heat pipe catheters. These control mechanisms include passive techniques such as gas loading, excess liquid charging, vapor flow modulation, and liquid flow modulation. These passive and/or active temperature control mechanisms allow temperature control to be within $\pm 0.02°C$. These techniques when coupled with an MHP catheter constructed from any one of a number of materials and working fluids could successfully be used for tissue cauterization in surgical applications or the destruction of cancerous tissues.

Fletcher and Peterson [66] claimed that MHPs are capable of providing an acceptable method for isolating and eliminating cancerous tissues in a number of applications where traditional surgical removal is not possible. These techniques may provide significantly improved treatment techniques for both operable and inoperable cancerous tissues.

Peterson and his coworkers are developing a tiny, highly efficient heat spreader to be used in a new device to be implanted in the brain of patients who suffer from severe epileptic seizures. The implant device is designed to detect and arrest epileptic seizures as they begin by cooling a small region of the brain, thereby effectively blocking the erratic electrical activity.

C. SPACECRAFT THERMAL CONTROL

Two new MHP concepts have been developed: wire-bonded heat pipe arrays with an effective conductivity of 30 times that of solid aluminum and flexible polymer heat pipes have been fabricated and modeled. The polymer heat pipes offer a greater degree of flexibility and a potentially higher effective thermal conductivity than any previously developed mechanism. Applications of these two concepts have a wide range of applications that extend well beyond spacecraft radiators [67]. A technique has been proposed for fabricating a series of MHPs into thin, flat, or curved panels made of refractory materials. These panels would exhibit a very high effective thermal conductivity and could be used as a highly conductive outer skin for the leading edge of engine cowling or wing surfaces on hypersonic vehicles. The resulting high effective conductivity of the exterior skin would serve to spread and dissipate the heat generated by local stagnation heating on these surfaces [68,69].

D. SOLAR ENERGY CONVERSION

Baker *et al.* [70] at NASA described a heat pipe solar dynamic space power receiver. Andraka *et al.* [71] and Adkins [72] reported use of an alkali metal heat pipe solar receiver to transfer thermal energy from the focus of a concentrator to the heater tubes of a dish–Stirling system.

E. AIRCRAFT TEMPERATURE CONTROL

Exterior surfaces on hypersonic aircrafts experience intense stagnation heating rates during ascent and reentry portions of their trajectories. Liquid metal heat pipes might reduce the temperature of leading-edge surfaces on the wings and engine cowl. Boman and Elias [73] tested a Hastelloy X-sodium heat for use on the wing leading edge of an advanced reentry vehicle. Cao and Faghri [74,75] developed a finite-difference numerical model for simulating the transient performance of nose cap and wing leading edge heat pipes. Heat pipes have also been considered for cooling engine components in conventional aircrafts. Gottschlich and Meininger [76] described the cooling of gas turbine engine vanes using heat pipes.

F. FUEL CELL TEMPERATURE REGULATION

A gas reservoir regulated heat pipe system is being used to provide the passive temperature regulation of a reformer and fuel cell system. Faghri's invention [77,78] is directed to a system and method for distributing heat in a fuel cell stack through bipolar interconnection plates, having one or more heat pipes disposed within the plate. His invention is directed to bipolar interconnection plates that distribute heat more effectively through the use of heat pipes disposed within the plate itself.

III. Steady-State Operation of Micro Heat Pipes

A. LITERATURE REVIEW OF VARIOUS STEADY-STATE MODELS

The steady-state models for fluid flow and capillary limitation of a system of MHP of any geometry and inclination is relevant for the complete understanding of the transport processes and the design of an MHP.

Cotter [48] introduced the concept of very small MHPs incorporated into semiconductor devices to promote more uniform temperature distribution and to improve thermal control. The first steady-state model was given by Cotter in 1984 to calculate the capillary limit for evaluating the performance of micro heat pipes. There were a few assumptions made by Cotter. The local axial heat transport $H(L)$ was assumed beforehand to predict the maximum transport capacity. Zero liquid–vapor interface velocity was assumed and the vapor was assumed to be in laminar flow and incompressible [48]. The expression for the maximum heat transport is given as

$$Q_{\max} = \left(\frac{0.16\beta\sqrt{K_l^+ K_v^-}}{8\pi H(l)}\right)\left(\frac{\sigma_l \lambda_l}{v_l}\right)\left(\frac{v_l}{v_v}\right)^{\frac{1}{2}}\left(\frac{A_t^{0.75}}{L}\right) \quad (1)$$

where $H(l)$ is the integral of axial heat transport fraction over L, K_l^+ and K_v^- are the flow shape factors, β is the geometric factor, L is the length of the pipe, A_t is the total cross-sectional area of the pipe, σ_l is the surface tension, λ_l is the latent heat of vaporization, v_l is the specific volume of liquid, and v_v is the specific volume of gas.

Babin et al. [17] developed a steady-state model for a trapezoidal MHP and the effects of the extremely small characteristic dimensions on the conventional steady-state heat pipe modeling techniques were examined. They used techniques outlined by Chi [79] and Dunn and Reay [9] to determine five different limits of sonic, entrainment, boiling, viscous, and capillary. These five limits govern the maximum heat transport capacity of a heat pipe. However, they assumed the profile of the liquid film to be known for calculating the liquid and vapor pressure drops. The mathematical form is

$$\Delta P_{\text{capillary}} \geq \Delta P_+ + \Delta P_\| + \Delta P_l + \Delta P_v \quad (2)$$

The results were significantly different compared to the model of Cotter [48] which was as a result of different assumptions made in both models. The model predicted dry out with a reasonable degree of accuracy between operating temperatures of 40°C and 60°C. However, above 60°C the model is slightly under predicted dry out.

Gerner *et al.* [80] compared the models of Cotter [48] and Babin *et al.* [17]. Contribution of the model by Babin *et al.* [17] was inclusion of gravity and recognition of vapor pressure loss. However, the assumption that the pressure gradient in the liquid flow passages was similar to that occurring in Hagen–Poiseuille flow was questionable. A scaling argument for the liquid pressure drop was presented and the average film thickness was estimated to be approximately one-fourth of the hydraulic radius. This resulted in a modification to the capillary limit. They also hypothesized that capillary limit may never be reached due to Kelvin–Helmholtz-type instabilities. Gerner *et al.* [80] included an interfacial stress at liquid–vapor interface. Neglecting this leads to an overestimation of maximum heat transfer capacity. They estimated the average film thickness to be approximately one-fourth the hydraulic radius. They calculated maximum heat flux by using the capillary pressure limit. The final result becomes

$$Q_{max} = \frac{3\pi}{2048} \frac{\rho_l \lambda_l D^3}{V_v L} \qquad (3)$$

where ρ is the density, λ is the latent heat of vaporization, and D is the diameter of the vapor path. The quantity v is the vapor kinetic viscosity and L is the length of the MHP.

Longtin *et al.* [18,81,82] developed one-dimensional (1D) mass, momentum, and energy equations for both liquid and vapor. Their model has only evaporator and adiabatic sections and they have solved them numerically to yield pressure, velocity, and film thickness information along the length of the pipe. Interfacial and vapor shear stresses were also included in the model. Apart from that, the convection and body force (gravity) terms have been included in the momentum equation. Maximum heat transport capacity was obtained which varied with the inverse of length and the cube of its hydraulic diameter. Profiles for pressure, velocity, and radius of curvature along the length were obtained. Effect of variation in operating parameters on the performance of the MHP was studied. There was good agreement with experimental results of Babin *et al.* [17]. The correct trend of increasing heat transfer with increasing operating temperature was observed.

Khrustalev and Faghri [82] developed a mathematical model to examine heat and mass transfer processes in an MHP. It was a numerical model to obtain the maximum heat transfer capacity (Q_{max}) and the overall thermal resistance. The model described the distribution of the liquid in an MHP and its thermal characteristics depending upon the liquid charge and the applied heat load. A relation for calculating the amount of the working fluid was given. The liquid flow in the triangular-shaped corners of an

MHP with a polygonal cross section is considered by accounting for the variation in the curvature of the free liquid surface and the interfacial shear stresses due to a liquid vapor frictional interaction. The mean internal heat transfer coefficients between the wall and the vapor were determined, which were used to calculate the overall thermal resistance of the condenser. A variation in the contact angle of the meniscus along the condenser was also considered. The disjoining pressure was considered for the evaporative section. Expression for maximum thermal resistance of the evaporator was also given. The radius of curvature decreased monotonically along the axial direction. The pressure drop in the liquid was several times higher than that of vapor and it increased significantly in the evaporator as the cross sectional area of the liquid was small. The maximum heat transfer increased with an increase in the liquid charge whereas it decreased with an increase in the minimum contact angle at the evaporative end. The results were compared with the experimental data by Wu and Peterson [19]. There was good agreement between the experimental results of the onset of dry out and the presented model. There was an overestimation of the maximum heat transfer capacity when the shear stress at the free surface of the liquid due to a liquid–vapor frictional interaction was neglected. Khrustalev and Faghri [83] considered the thermal resistances in condensation and evaporation parts of an MHP with triangular grooves. Most of researches have, however, focused on the analysis of two-dimensional (2D) steady-state heat transport capacity of MHPs and the local variation of shear stress at the liquid–vapor interface is neglected.

Wang *et al.* [84] have used the porous medium concept to build up a fundamental model for two-phase flows inside mini channels. The capillary force, which is important for MHPs, is rigorously included in their model by following the same treatment as for porous media. They have demonstrated adiabatic co- and countercurrent flows and heated forced and natural convective flows. They compared some fundamental results such as the two-phase pressure drop in a down flow, the flooding limit in a countercurrent flow, and dry-out heat flux in natural convection boiling, from previous studies available in the literature. With main emphasis on the applications of micro heat pipes, their model is used to predict the capillary limit of an operating MHP, to explain the liquid holdup phenomenon and to imply the onset and origin of the plug flow pattern.

Peterson and Ma [20] developed a mathematical model to predict minimum meniscus radius and the maximum heat transport in triangular grooves. A method for determining the theoretical minimum meniscus radius was developed and used to calculate the capillary heat transport limit based on physical characteristics and geometry of the capillary grooves. Variations in both friction factor and the cross-sectional area of the liquid flow were considered. The model assumed an optimum amount of working fluid and neglected the effect of condensing heat transfer coefficient. The results showed that with an increase in the contact angle, the heat transport capacity of the MHP initially increased, reached a maximum and finally decreased. This implied that there is an optimum contact angle that resulted in maximum heat transport capacity. Tilt angles were also shown to have a significant effect on the capillary heat transport. Heat transport capacity decreased as the length of the heat pipe increased. A decrease in the apex angle increased the heat transport capacity. There was good agreement with the experimental results of Ma and Peterson [24] for the maximum heat transport in triangular grooves. This verified that there is an optimum dimension for which the capillary grooves have the largest capillary heat transport capability.

Ha and Peterson [26] predicted the axial flow of evaporative thin films through a V-shaped microchannel for very small tilt angles. The addition of the gravity term due to the tilt angle changed the governing equations from linear to nonlinear. As the effect of the tilt angle is small, a perturbation method was applied to obtain a first-order perturbation solution of the radius of curvature. A perturbation parameter was expressed as a function of five parameters which were inclination angle, length of adiabatic region, Bond number, capillary number, and groove shape. The perturbation parameter was inversely proportional to the square root of Bond number. Bond number and capillary number have dominant effect for a fixed geometry. This perturbation model can be applied even for large tilt angles if the heat input is large enough.

Peterson and Ma [85,86] presented a detailed mathematical model for predicting the heat transport capability and temperature gradient that contribute to the overall axial temperature drop as a function of heat transfer in an MHP. The model utilized a third-order ordinary differential equation which governed the fluid flow and heat transfer in the evaporating thin film region. With this model, the temperature distribution along the axial direction of the heat pipe and the effect on the heat transfer was predicted. An experimental investigation was also conducted in order to verify the model presented and a comparison with experimental data was done. The comparison indicated excellent correlation between the analytical model and experimental results. The analysis provides a better understanding of the heat transfer capability and temperature variations occurring in MHPs.

Sartre *et al.* [87] used a three-dimensional (3D) steady-state model for predicting heat transfer in an MHP array. Three coupled models to solve the microregion equations, the 2D wall heat conduction problem and the longitudinal capillary two-phase flow were developed. The results, presented for an aluminum/ammonia triangular MHP array, showed that the major part of the total heat input in the evaporator section goes through the microregion. In addition, both the apparent contact angle and the heat transfer rate in the micro region increase with increasing wall superheat. It was also shown that the inner wall heat flux, temperature and the contact angle decrease along the evaporator section.

Tio et al. [88] presented a 1D model of an MHP to investigate its thermal and fluid–mechanical behaviors within its capillary limitation. They used Darcy's equation for two-phase flows in porous media and Laplace's capillary equation for pressure drop. The effects of various parameters were incorporated into the analysis. These parameters were the capillary number, the charge level of the working fluid, the liquid–vapor viscosity ratio, contact angle, the relative lengths of the evaporator and the condenser sections, the orientation of the MHP, and the Bond number. Furthermore, comparison with existing experimental results shows that the porous-medium model is reasonably adequate for the prediction of the capillary performance limit of an MHP.

Catton and Stroes [28] developed a 1D semianalytical model to predict dry-out length for inclined triangular capillary grooves subject to heating from below. The model utilizes a macroscopic approach and employs the concept of an apparent contact angle. The concept of accommodation theory was introduced to account for a change in the radius of curvature of the liquid–vapor interface between the liquid and the vapor reservoirs. This meant that the slope of the groove walls in combination with a zero macroscopic contact angle forces the liquid–vapor interface to take on a particular radius of curvature compatible with the local groove geometry. Accommodation length was the axial distance over which this transformation occurs. The apparent contact angle till the accommodation length was greater than zero and was varying. There was good agreement with the experimental results. A design curve was also developed to estimate dry-out length. However, adiabatic and condenser sections were not considered in this study.

Riffat et al. [89] developed an analytical model to investigate the performance of a miniature gravitational heat pipe and two microgravitational heat pipes of different sizes using water as the refrigerant. The model determined the limits of heat transport capacity as functions of properties of working fluid, inclination angle, and liquid fill level. The limits of heat pipes express the thermal performance of heat pipes. The entrainment and capillary limits were found to be the dominant limits for an MHP. The capillary limit was found to influence the heat transport performance for operating temperature of $<80°C$. MHP with smaller width showed better thermal performance than the MHP with larger width. The critical limit increased with an increase in the inclination till $30°$ and after that it remained constant. A numerical model was also developed to investigate the operating characteristics of heat pipes. The expression for the free molecular flow mass flux of evaporation presented by Collier [90] and later by Colwell and Chang [91] was used in this model.

Cross-sectional areas, pressures, temperatures, and mass flow rates of the liquid and vapor within the heat pipes were presented.

Zhang and Wong [92] observed lower effective thermal conductivity of an MHP pipe as compared to conventional heat pipes reasoning the complexity of the coupled heat and mass transport, and the complicated 3D bubble geometry inside MHPs, there is a lack of rigorous analysis. As a result, the relatively low effective thermal conductivity remains unexplained. An idealized MHP was conceptualized, that eliminates the complicated geometry, but retained the essential physics. They studied many effects such as thermocapillary flow and evaporation and condensation physics using a simple geometry. The flow field induced by evaporation was also presented.

Anand et al. [27] developed a numerical model to predict the onset, location, and propagation of the dry-out point for large tilt angles. They modified the general model developed by Ha and Peterson [26] by incorporating the variation of evaporative heat flux along the length of the groove. The heat losses through convection from top, bottom, and sides of the substrates were also taken into account for calculating the evaporative heat flux. There was good agreement between the predicted values and the experimental results. Dry region propagated away from the heater with an increase in inclination and heat input to the system.

Kalahasti and Joshi [93] combined numerical and experimental investigation on a novel flat plate MHP spreader to better understand the effect of primary operating parameters governing the performance of such devices. A numerical thermal model was developed to predict the temperature response with variation in the leading geometrical, material, and boundary parameters of the spreader, namely, wall thickness, thermal conductivity, power input, and heat source size. The results showed that, unlike conventional heat pipes, wall thermal conductivity was a major factor in such thin, flat spreaders. The spreader performance also degrades with a decrease in a heat source size. Visualization experiments were conducted to qualitatively understand the heat transfer phenomena taking place on these devices. These confirmed that the primary limitation to heat transfer from these devices was due to the capillary limitation of the wick structures.

Kim et al. [94] developed an analytical model for heat and mass transfer in a miniature pipe with a grooved wick structure to give a maximum heat transport rate and the overall thermal resistance under steady-state condition. The effects of liquid–vapor interfacial shear stress, contact angle, and amount of initial liquid charge were considered in the model. A modified Shah method was suggested to account for the effect of the liquid–vapor interfacial shear stress on the thermal performance of a heat pipe. The proposed model was verified by conducting experiments for measuring the maximum heat transport rate and the thermal resistance. There was good agreement between the results from the proposed study and the experimental results. The results showed that the radius of curvature increases nonlinearly along the axial direction toward the condenser and it increases rapidly at the beginning of the condenser section. Numerical optimization was performed to enhance thermal performance of the miniature heat pipes.

In context of a European Commission-funded project, Khandekar et al. [95] studied the development of a standardized multifunctional stacked 3D package for potential applications in aviation, space, and telecommunication sectors. They aimed that the standardization and modularity would integrate packages from different technologies and will allow mutual slice interchangeability. Three potential options were studied, that is, (1) module liquid cooling, (2) integration of miniature copper–water cylindrical heat pipes (OD = 3.0 mm) with the 1.0-mm substrate slice, and (3) development of flat plate heat pipes of 0.9-mm thickness. For options (1) and (2), initial tests have been performed taking aluminum as a representative material for AlSiC metal matrix composites which were to be employed in the final design. Further, copper-based flat plate microstructure conventional heat pipes have been developed and performance tested. Thermal interactions have been investigated with thermocouples coupled with IR thermogram. For a safe operation up to 30-W heating power (10 W per slice), while thermal diffusion through the bare metallic substrate is sufficient for heat transfer from chip to the substrate, MHPs should be employed to cool the substrate and transfer heat from it to an external cold plate.

Do et al. [96] investigated heat transfer and fluid flow characteristics in an MHP with curved triangular grooves using numerical and experimental methods. In the numerical part, a 1D mathematical model for MHPs with curved triangular grooves was developed and solved to obtain the maximum heat transport rate, the capillary radius distribution, and the liquid and the vapor pressure distributions along the axial direction of the MHP under the steady-state condition. In particular, the modified Shah method was applied to calculate the pressure drop induced by the liquid–vapor interfacial shear stress.

Suh and Park [97] presented the model of an MHP based on the analysis by Chi [8] in a steady-state operation. The thermal performance of an MHP in an axial flat grooved channel was investigated numerically and the interfacial shear stress caused by the velocity difference between the liquid and the vapor was considered. The MHP studied in their work used ammonia as the working fluid. The results were obtained primarily in the operation temperature of 300 K, which are the flows of liquid and vapor in trapezoidal grooves, the axial variation of pressure difference between vapor and liquid, contact angle, velocity of liquid and vapor, and so forth. In addition, the maximum heat transport capacity of MHP was obtained by varying the operational temperature and was successfully compared with the result from Schneider and Devos's model [98] in which the interfacial shear stress was neglected.

Sheu *et al.* [99] analyzed the capillary performance of triangular microgrooves. The influences of several major parameters on the capillary performance (wetted axial length) of triangular microgrooves were discussed theoretically. One-dimensional nonlinear and contact angle possessed differential and algebraic equations are used for the theoretical analyses of triangular microgrooves. The curvature radius, cross-sectional area, and distribution of pressure and velocity of the working fluid in the microgrooves were considered. Besides, the mutual effects among inertial force, body force, capillary force, and friction force were also discussed. The significance of contact angle and hydraulic diameter for the prediction of capillary performance of microgrooves are demonstrated by the proposed algebraic solutions.

Suman *et al.* [29] presented a numerical model which was based on the first principles utilizing a macroscopic approach. The coupled nonlinear governing equations for the fluid flow, heat and mass transfer were developed and solved numerically. All three sections, that is, evaporative, adiabatic, and condensing sections were considered in the study. Profiles of radius of curvature, liquid velocity and liquid pressure were generated. The profile of the radius of curvature was used to predict the onset of the dry-out point and the propagation of the dry-out length. A method was developed to estimate the dry-out length as a function of system geometry and process variables, e.g., heat input and inclination. The results of the model were successfully compared with the experimental results from the previous study by Anand *et al.* [27]. Later, this model was modified by Suman and Hoda [31] by changing the boundary conditions such that the evaluation of dry-out length became easier, and considered a substrate temperature profile in their model.

Suman and Kumar [32] presented an analytical model for fluid flow and heat transfer in an MHP of a polygonal shape by utilizing a macroscopic approach. The coupled nonlinear governing equations for fluid flow, heat and mass transfer were modified and solved analytically. The analytical model was able to study the performance and the limitations of MHP. The model provided analytical expressions for the critical heat input, dry-out length, and available capillary head for the flow of fluid. A dimensionless parameter, which predicts the performance of an MHP, was also obtained from this model. The results predicted by the model were compared with the published results in literature [18,27,89] and good agreement was obtained.

1. A Typical Steady-State Model

A typical steady-state model for an MHP, which is the modified version of the presented by Suman and his coworkers [29,31], is taken. A V-shaped MHP has been considered here. An optimal amount of the coolant liquid is charged to the system so that at steady state, the cold end remains filled with the coolant liquid such that the radius of curvature can be calculated from the geometry of the system ($R = R_o$). The model presents equations for all the three sections, that is, evaporative, adiabatic, and condenser encompassing the complete microscopic heat pipe. They are under the following assumptions:

(1) The 1D steady incompressible flow along the length of a heat pipe since the flow along the transition region is small [23]; (2) negligible heat dissipation due to viscosity; (3) 1D temperature variation along the length of a heat pipe; (4) negligible shear stress at the liquid–vapor interface because the channel area for vapor flow is relatively large and at the low heat fluxes, the vapor velocity is generally small. This justifies the assumption of no shear at the liquid–vapor interface [20,28]. (5) Predefined heat flux distribution (Q) with position used by the coolant liquid. The fluid flow is governed by the pressure difference between the hot and the cold ends. The vapor pressure affects evaporation and condensation. However, a successful formulation of dependency of heat fluxes on the vapor pressure has not been done. Therefore, the predefined heat input has been taken [19,26]. (6) Convective heat loss has been neglected since the two-phase heat transfer is orders of magnitude higher than the natural convection. (7) Disjoining pressure has not been included and (8) constant pressure in the vapor region. The constant pressure in the vapor region assumption is valid in this case as the vapor flow space in the channels is quite large especially considering low heat fluxes. The vapor pressure drop required for the flow has been calculated and found to be very small.

One corner of a section of a heat pipe of any polygonal shape of length Δx is shown in Fig. 3.

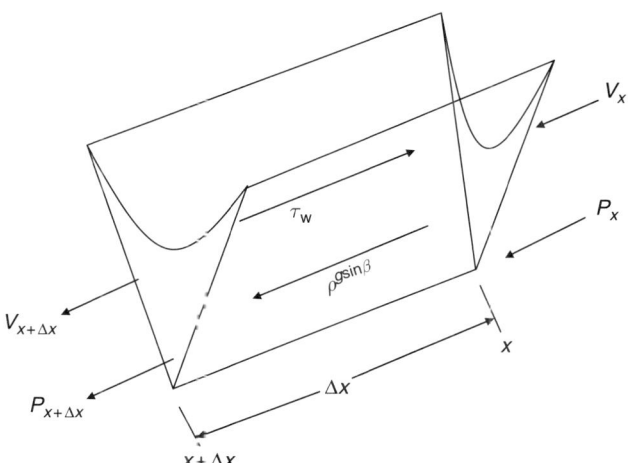

FIG. 3. One corner of a section of a heat pipe of a polygonal shape.

The liquid pressure as a function of the radius of curvature can be estimated from the Young–Laplace equation in the differential form as

$$\frac{dP_l}{dx} = \frac{\sigma_l}{R^2}\frac{dR}{dx} \qquad (4)$$

The steady-state momentum balance in the differential form is given as

$$\rho_l A_l V_l \frac{dV_l}{dx} + A_l \frac{dP_l}{dx} + 2L_h \tau_w + \tau_{l,i} R_m - \rho_l g \sin\beta A_l - \frac{A' A_l}{2\pi h^4}\frac{dh}{dx} = 0 \qquad (5)$$

This is the modified equation of the published model by Suman and coworkers. The modified terms are fourth and sixth. The fourth term is for the shear stress at the interface and the sixth term is for the disjoining pressure. The disjoining pressure has Hamaker constant (A') and the liquid height (h) [34]. The liquid height is varying across the cross section. But the liquid height at the apex has been considered since our analysis mainly focuses on the capillary region. The substantial reduction in the liquid pool thickness is in the transition of the capillary and the thin-film regions.

The differential form of the steady mass balance for the coolant liquid is given as

$$\frac{d(\rho_l A_l V_l)}{dx} + \frac{Q_v R_m}{\lambda_l} = 0 \qquad (6)$$

The differential form of the steady state mass balance for the vapor is given as

$$\frac{d(\rho_g A_g V_g)}{dx} - \frac{Q_v R_m}{\lambda_l} = 0 \qquad (7)$$

The energy balance equation for the coolant liquid, after considering sensible heat change in the volume element, is given as

$$\rho_l C_{pl} V_l A_l \frac{dT_l}{dx} = QW_b - Q_v R_l \qquad (8)$$

where Q is the input heat flux taken by the coolant liquid from the substrate. The quantity Q is positive when added (in the evaporative section) and is negative when extracted (in the condenser section).

The steady-state energy balance in the substrate is given as

$$A_{cs} K_s \frac{d^2 T_s}{dx^2} - QW_b = 0 \qquad (9)$$

where the first term is the net change in the conductive heat in the control volume and the second term is the heat taken up by the coolant liquid.

a. Nondimensionalization. The following nondimensional and associated parameters have been used: friction factor $(f) = K'/N_{Re}$, Reynolds number $(N_{Re}) = (D_h \rho_1 V_1)/\mu_1$, hydraulic diameter $(D_h) = 4A_1/2L_h$, wall shear stress $(\tau_w) = (\rho_1 V_1^2 f)/2$, shear at the liquid–vapor interface $(\tau_{l,i}) = [\rho_g(V_g - V_1)f/2] - \rho_g V_g(V_{g,i} - V_1)$ (the velocity in the bracket is the relative velocity of the vapor), reference velocity $(V_R) = Q'/\rho_1 R_o^2 \lambda_1$, reference pressure $(P_R) = \sigma_1/R_o$, reference height

$$h_R = R_o \left[\frac{\cos(\alpha + \gamma)}{\tan \alpha} + \sin(\alpha + \gamma) - 1 \right]$$

and reference temperature $(T_R) = T_{con}$.

The dimensionless parameters are defined as follows: dimensionless radius of curvature $(R^*) = R/R_o$, dimensionless position $(X^*) = x/L$, dimensionless liquid velocity $(V_1^*) = V_1/V_R$, dimensionless vapor velocity $(V_g^*) = V_g/V_R$, dimensionless liquid pressure $(P_1^*) = P_1/P_R$, dimensionless substrate temperature $(T_s^*) = T_s/T_R$, dimensionless height $(h^*) = h/h_o = R^*$. The quantity K' is used in the expression of friction factor (f) and is a constant for a specific geometry [100,101]. The quantity R_o is the radius of curvature at the cold end and is a function of side length, the contact angle for the substrate and coolant liquid system and the groove angle. The final set of nondimensionalized equations is as follows:

$$\frac{\tau g \sin(\beta)}{V_R} + \frac{\tau Q_v R_m V_1^*}{A_1 \rho_1 \lambda_1} - \frac{\tau B_2 V_1^*}{\rho_1 (R_o R^*)^2} - \frac{\tau \tau_{l,i} R_m}{A_1 \lambda_1}$$
$$+ \left[\frac{2\tau V_R V_1^{*2}}{LR^*} - \frac{\tau \sigma_1}{\rho_1 V_R R_o LR^{*2}} + \frac{\tau A'}{2\pi \rho_1 V_R L h_o^3 R^{*4}} \right] \frac{dR^*}{dX^*} = 0 \quad (10)$$

$$\frac{\tau R^* Q_v R_m L}{2\rho_1 A_1 \lambda_1} + \frac{2\tau V_R V_1^*}{LR^*} \frac{\partial R^*}{\partial X^*} + \frac{\tau R^* V_R}{2L} \frac{dV_1^*}{dX^*} = 0 \quad (11)$$

$$\frac{dP_1^*}{dX^*} = \frac{1}{R^{*2}} \frac{dR^*}{dX^*} \quad (12)$$

$$\frac{dV_g^*}{dX^*} = -\frac{Q_v R_1 L}{\rho_1 A_g \lambda_1 V_R} + \frac{2A_1 V_g^*}{A_g R^*} \frac{dR^*}{dX^*} \quad (13)$$

$$Q_v = \frac{1}{R_1} \left[Q W_b - \frac{\rho_1 C_{pl} V_R A_1 V_1^*}{L} \frac{dT_1}{dX^*} \right] \quad (14)$$

$$\frac{d^2 T_s^*}{dX^{*2}} - \frac{QW_b L^2}{T_R A_{cs} K_s} = 0 \tag{15}$$

$$Q = \frac{Q'}{fW_b L} \tag{16}$$

Equations (10)–(16) are valid for all three sections of a microgrooved heat pipe, namely, evaporative, adiabatic, and condenser. The quantity Q is zero in the adiabatic section, negative in the condenser section (heat is extracted), and is positive in the evaporative section (heat is supplied). These equations can be solved numerically taking predefined heat interaction flux between the solid substrate and the coolant liquid is Q.

Here geometrical parameters A_1, W_b, R_o, and B_2 are taken from Suman and Hoda [31], considering that the ungrooved area is zero. They are given as

$$A_1 = R^2 \left[\{\cot(\alpha + \gamma) - \phi/2\} + \frac{\cot(\alpha + \gamma)\cos(\alpha + \gamma)\sin \gamma}{\sin \alpha} \right] \tag{17}$$

$$A_g = A_{total} - A_1 \tag{18}$$

$$W_b = 2a \tag{19}$$

$$R_o = \frac{a \sin \alpha}{\cos(\alpha + \gamma)} \tag{20}$$

$$B_2 = \frac{\mu_1 K' \cos^2(\alpha + \gamma)}{2 \sin^2 \alpha \left[\frac{\cot(\alpha + \gamma)\cos(\alpha + \gamma)\sin \gamma}{\sin \alpha} + \{\cot(\alpha + \gamma) - \phi/2\} \right]^2} \tag{21}$$

The dimensionless boundary conditions can be written as

at the cold end: $(X^* = 1)$, $R^* = 1$, $P_1^* = (P_{vo}/P_R) - 1$, $V_1^* = 0$, $T_s^* = T_{con}/T_R$,

at the hot end: $(X^* = 0)$, $Q_{heater} = -\frac{K_s A_{cs} T_R}{L} \frac{\partial T_s^*}{\partial X^*} \bigg|_{X^* = 0}$

b. Results and Discussion. To present a typical MHP steady-state model results, a V-shaped heat pipe has been considered. The silicon substrate is 0.8 cm wide and 2.0 cm long. The length of the evaporative, adiabatic and condenser sections have been assumed to be equal. The V-groove width and groove spacing are 0.1 mm and the apex angle of groove is 60°. A set of 10 such grooves has been considered herein. The temperature at the condenser end and the reference temperature have been considered as 32°C. The

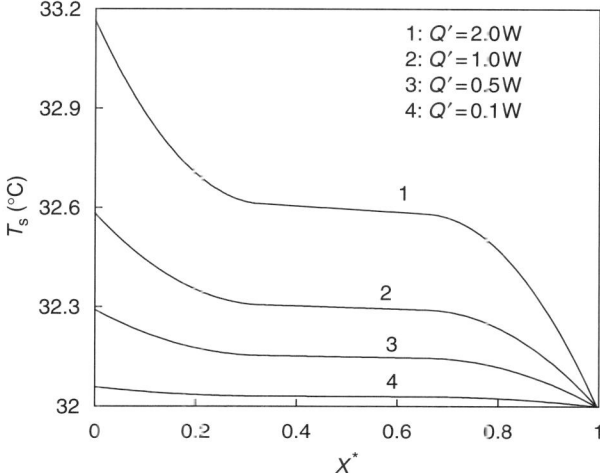

FIG. 4. Substrate temperature profiles (T_s) as a function of dimensionless axial position for different heat inputs (Q').

inclination of the heat pipe is 10°. A constant heat flux distribution has been considered. Pentane has been taken as the working fluid and silicon as the substrate. The contact angle is considered as zero, although the presented model is capable of handling nonzero contact angle.

The substrate temperature profiles as a function of the dimensionless axial position for different heat inputs are presented in Fig. 4. It has been seen that the temperature difference along the heat pipe decreases with a decrease in the heat input. Therefore, it is claimed that the heat pipe can remove heat without generating much temperature difference and this property of a heat pipe makes it very promising for many potential applications.

The profiles of the dimensionless radius of curvature as a function of the dimensionless axial position of the heat pipe (R^* vs. X^*) for different heat inputs are presented in Fig. 5. Figure 5 shows the gradual decrease in R^*, as the hot end is approached, signifying higher curvature at the hot end. The quantity R^* at any location also decreases with an increase in the heat input. The behavior of the radius of curvature gives a qualitative idea of the capillary pumping. The value of R^* at the hot end or the gradient of R^* can also predict the operating limit of a microgrooved heat pipe, that is, the critical heat input for a system controlled by capillary pumping.

The variation of the axial liquid velocity along the dimensionless axial position is presented in Fig. 6. The direction of velocity is from the cold end

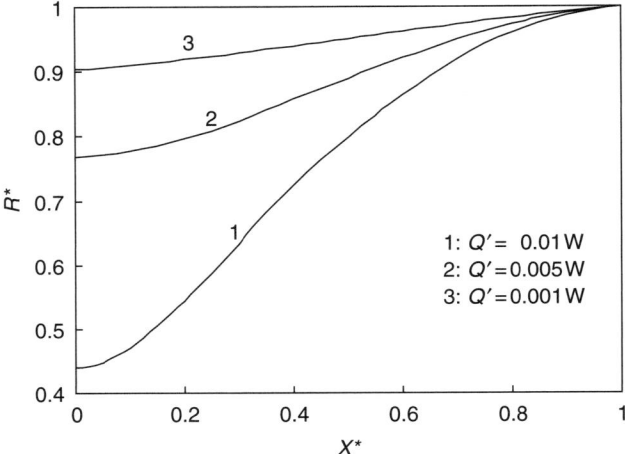

Fig. 5. Dimensionless radius of curvature as a function of dimensionless axial position for different heat inputs (Q').

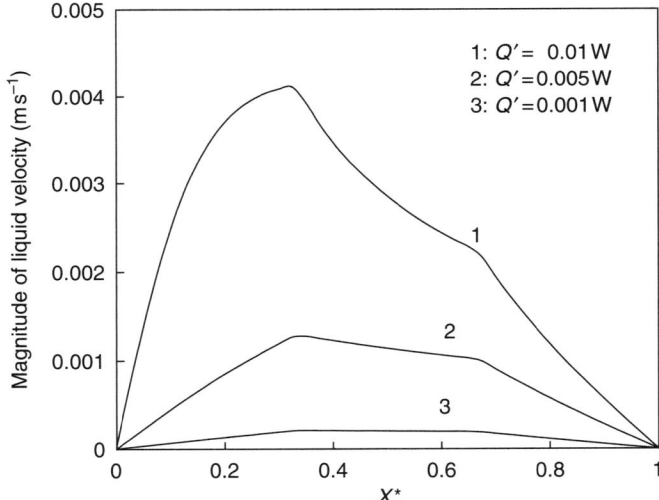

Fig. 6. Variation of axial liquid velocity along the dimensionless axial position.

to the hot end. The absolute value of the axial liquid velocity is zero at the hot end and increases in the evaporative region. This is due to the cumulative effect of replenishing the amount evaporated throughout the evaporator region. Although there is no evaporation and condensation in the adiabatic region, the liquid velocity decreases slightly. This is because of a slight increase in the radius of curvature in the adiabatic section (and hence an increase in the liquid flow area) and the friction loss. In the condenser region, there is a further decrease in the liquid velocity because condensation results in a sharper increase in the value of radius of curvature. Finally, at the end of the condenser region, the liquid velocity becomes zero. It can also be observed from Fig. 6 that with an increase in heat input, evaporation increases to replenish the enhanced amount of evaporated liquid. As a result the liquid velocity increases at any fixed location.

Total mass flow rate (flow rate of liquid + flow rate of vapor) across any cross section of a heat pipe should be zero. This equation gives the vapor velocity in a micro heat pipe. It is very important to calculate to check the sonic limitation, which will be addressed later. The variation of the axial vapor velocity along the dimensionless axial position is presented in Fig. 7. The direction of the vapor velocity is from the hot end to the cold end. It has a similar trend to that of the magnitude of the liquid velocity in the evaporative and the condenser sections. In the adiabatic section, the radius of curvature increases and the area available for the liquid flow decreases. Therefore, a slight increase in the vapor velocity has been obtained. But, they are opposite in direction. The magnitude of the vapor velocity is more than that of the liquid velocity. This is because of a very low density of the vapor as compared to the liquid.

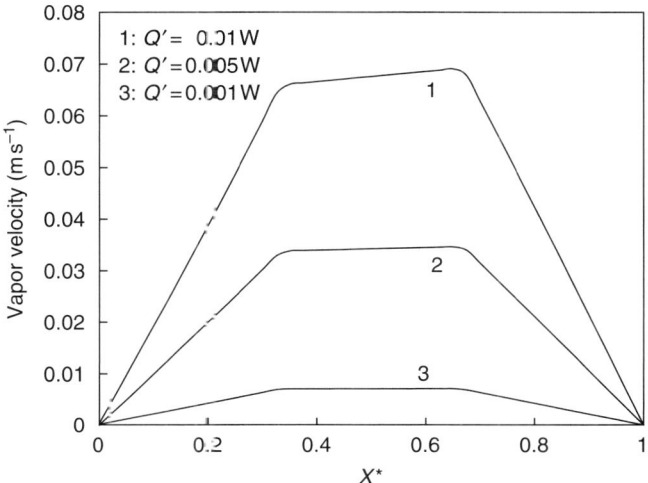

FIG. 7. Variation of axial vapor velocity along the dimensionless axial position.

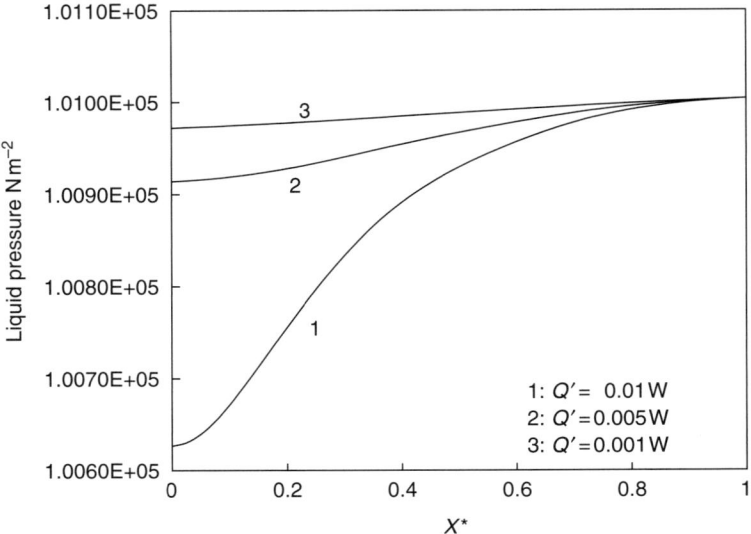

FIG. 8. Variation of liquid pressure along the dimensionless axial position.

In the microgrooved heat pipe, the liquid should flow from the cold end to the hot end. This requires the liquid pressure to decrease from the cold end to the hot end (dP^*/dX^* is positive for coordinate system used here). The variation of the liquid pressure along the dimensionless axial position is presented in Fig. 8. With an increase in the heat input, more liquid gets evaporated in the evaporative section. Therefore, the requirement of more liquid in the evaporative section demands the higher pressure drop between the cold end and the hot end to sustain an increase in fluid flow.

A part of the total heat supplied to the heat pipe is used by the liquid to raise its temperature while the remaining is used for evaporation. Only a small fraction of the heat input is used to increase the sensible heat of the liquid. It has been found that the amount of heat used as sensible heat is negligible compared to the heat used for the evaporation and the condensation. Hence, the sensible heat has negligible effect on the performance of a heat pipe. The heat flux Q_v, used for the evaporation of the coolant liquid and the condensation of the vapor along the dimensionless axial position for a constant heat flux distribution, is presented in Fig. 9. In the evaporative section, the evaporation of the coolant liquid takes place resulting in Q_v having a positive value. No evaporation and condensation take place in the adiabatic region, and therefore Q_v is zero. In the condenser region, the condensation of the coolant liquid takes place resulting Q_v being negative.

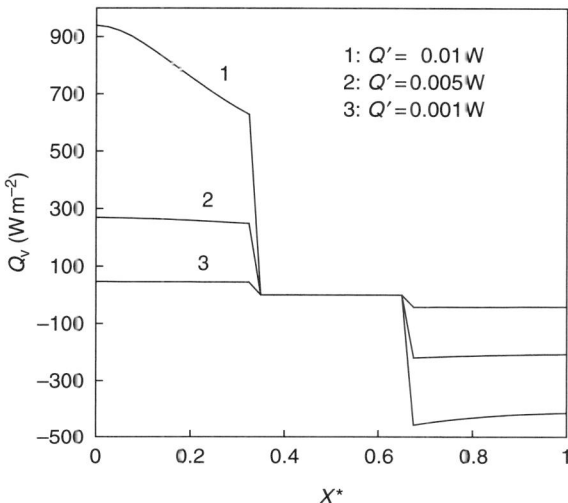

FIG. 9. Heat flux Q_v used for the evaporation of liquid along the dimensionless axial position.

B. STEADY-STATE EXPERIMENTAL STUDIES

Murakami et al. [102] performed one of the first experimental investigations by studying a flat heat pipe several millimeters thick. Two types of heat pipes were evaluated. The first one was a conventional flat heat pipe using a series of rectangular grooves machined into the upper and lower plate as the wicking structure. The second one was a series of triangular grooves machined into the plate to form an array of MHPs. A very limited amount of data was obtained for the flat heat pipe which used triangular MHPs. The results showed successful operation of an MHP using the sharp angled corner regions of a noncircular channel. The heat transport capacity was quite small as compared to a more conventional flat plate heat pipe. This was due to slugging occurring in the condenser region of the triangular heat pipes. Sato et al. [103] fabricated and tested the characteristic of an MHP (3 mm outside diameter) with 26 or 40 grooves (Cu and water). The heat-transfer coefficient of higher than 10,000 W $m^{-2} K^{-1}$, has been achieved, which can be used in electronic equipments.

Babin et al. [17] conducted an experimental investigation of several individual MHPs having approximately 1 mm of diameter. The purpose was to determine the accuracy of the previously described steady-state models, to verify the MHP concept and to determine the maximum heat transport capacity. The steady-state tests were conducted at inclinations which both assisted and hindered the return of a liquid to the evaporator. Heating in the

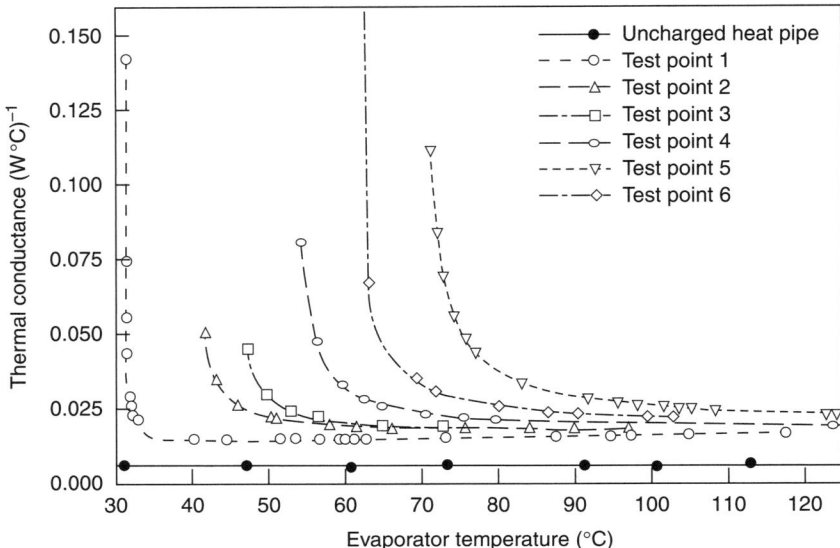

FIG. 10. Measured thermal conductance of a trapezoidal MHP as a function of the evaporator temperature taken from Babin *et al.* [16].

evaporator was done by the electrical resistance and a heat sink in the condenser was provided. The heat flux was incremented during the tests and the temperature of the coolant bath was adjusted to maintain a constant adiabatic wall temperature. Four test pipes were evaluated at a total of six adiabatic wall temperatures. The measured thermal conductance of a copper trapezoidal heat pipe as a function of the evaporator temperature is shown in Fig. 10. As dry out occurred, the thermal conductance of the charged copper heat pipe decreased rapidly with respect to the evaporator temperature and asymptotically approached a constant value slightly above than the values for an uncharged copper heat pipe. This result was expected as an increase in input power initially caused the liquid meniscus to recede into the liquid channel located in the corners which decreased the radius of curvature. A decrease in the radius of curvature decreased the cross-sectional area of the liquid. Since less area was available to transfer the heat, the evaporator temperature decreases. The steady-state model overpredicted the experimentally determined heat transport capacity at operating temperatures below 40°C and under predicted it at operating temperatures above 60°C.

Anand *et al.* [27] conducted experimental investigation to study the onset and propagation of the dry-out point on a chemically machined microgrooved surface on a silicon wafer with pentane as the coolant liquid. A computer-controlled mechanical surface profilometer was used to verify the apex angles and the dimensions. The axial temperature distribution was accurately measured with (wet) and without (dry) the working liquid as a function of the heat input and inclination of the substrate using small thermocouples. The results were compared with the temperature profiles of a nongrooved surface. The dry-out point was located by comparing the dry and the wet temperature profiles. It was shown that the relative cooling effect of the grooved silicon wafer was more than that of the nongrooved wafer. The ambient temperature played a critical role in the evaluation of the origin and propagation of the dry-out point.

Kim et al. [94] conducted an experimental investigation to verify their proposed mathematical model. They measured the maximum heat transport rate and thermal resistance. Their experimental apparatus comprised of the evaporator section, the adiabatic section, and the condenser section. A thin-film heater provided the uniform heat flux to the upper and lower heating blocks attached to the heat pipe. Three thermocouples were installed on the evaporator and the condenser section and one thermocouple was installed at the center of the adiabatic section. The working temperatures were chosen to be 40, 50, 60, and 70°C. The maximum error between the experimental data for the maximum heat transport rate and the numerical results was 3.8% which demonstrated the usefulness of the proposed model in the working range of 40–70°C.

Do et al. [96] conducted the experiments to validate the numerical model. In the experiments, the MHP with 0.56 mm in hydraulic diameter and with 50 mm in length tested. The experimental results for the maximum heat transport rate agreed well with those of the numerical investigations. Finally, the thermal optimization of the MHP with curved triangular grooves was performed using the numerical model.

Moon et al. [104] conducted experimental investigations with MHPs of triangular and rectangular cross sections to analyze their thermal characteristics. The material of the MHP was copper and the working fluid was pure water. The MHP was manufactured by the drawing process rather than the etching process. The testing apparatus shown in Fig. 11 is composed of an MHP, a vacuum chamber unit, a constant temperature bath for cooling, a data acquisition system, and a DC power supplying unit. The evaporator was heated using the electrical resistance heater and DC power supply unit. The condenser was cooled by a water jacket with circulating water. K-type thermocouples were installed by soldering at two points in the evaporator and condenser sections and one point in the adiabatic section. The experiments were performed in a vacuum chamber to minimize a heat loss. The operating temperature of the heat pipe was considered from 60 to 90°C. Results showed that the heat pipe with a triangular cross section can dissipate a thermal load of up to 7 W which was 1.6 times

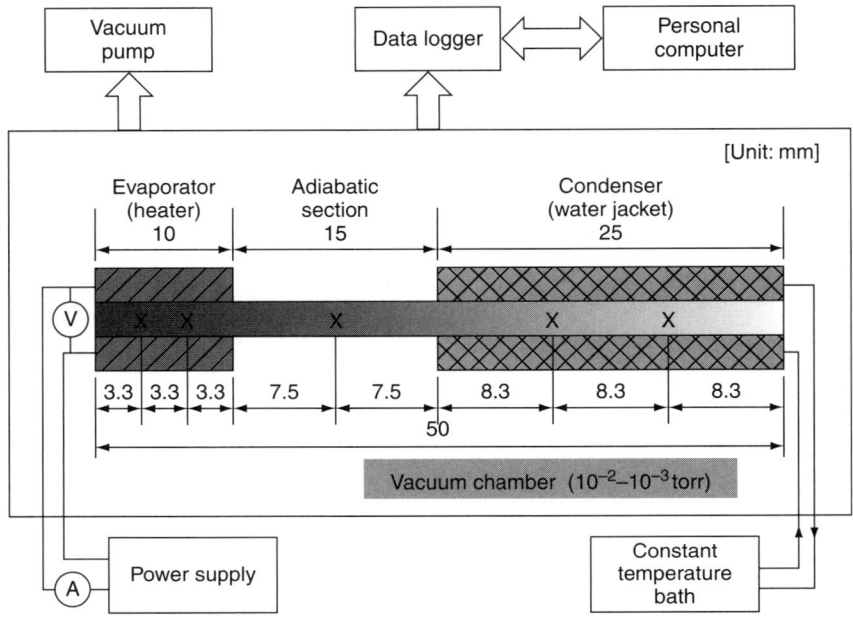

FIG. 11. Experimental apparatus for MHP study as taken from Moon et al. [87].

that of the MHP with a rectangular cross section. The heat transfer limit was a function of the operating temperate and it increased with an increase in the operating temperature. The temperature differences between the evaporator and the condenser were 4.3–9.8°C over the thermal loads of 0.5–4 W. The effect of the inclination angle was small and the thermal characteristics were more stable in the bottom heating mode than the top heating mode.

Zhou et al. [105] proposed the maximum applicable power of an MHP from the application point of view. They have presented the experimental relations among the maximum applicable power, air velocity, and wind temperature. The effect of using MHP for cooling modules was compared with that of using Cu fin and the comparison results indicated that the MHP is found to be advantageous.

Zhang et al. [106] studied the heat transfer characteristics of copper and silver MHPs. Under the condition of natural convection and vacuum, the experiment was done by way of electric resistance heater and water bath, respectively. The temperature performance and heat transfer capability as well as the effect of the tilt angles of the MHPs were studied and analyzed.

C. Electrohydrodynamically Augmented Micro Heat Pipe

In the past, several attempts have been made to increase the heat transport capacity of a heat pipe [107–113] since the MHP is a wickless heat pipe. As a result, its heat transport capacity is less than the wicked heat pipe. The electrohydrodynamically (EHD)-assisted heat pipe offers promise to improve the performance of an MHP. It provides an overall mass saving, which is a prime NASA concern [114]. Researchers have attempted to improve the heat transport of a heat pipe using EHD phenomena. Jones [115] proposed to replace a capillary wick structure in a heat pipe with an EHD pump that utilized polarization forces to generate pumping. Jones and Perry [116] used an EHD heat pipe successfully, but the performance was poorer than the existing capillary-driven heat pipe. Loehrke and Debs [117] improved the electrohydrodynamic (EHD) heat pipe of Jones and Perry, and were able to achieve equivalent thermal performance of conventional axial-groove heat pipes. Bologa and Savin [118] used the dielectrophoretic force to enhance the heat transport capacity in an experimental heat pipe operating as a two-phase thermosiphon, and obtained 53% increase in the heat transport capacity after using 36 kV. The enhancement of a heat pipe transport capacity utilizing the Coulomb force was investigated by Babin et al. [119]. They used an ion-drag pump to generate the Coulomb force, and to increase the capillary limit of a heat pipe. Sato et al. [120] proposed using the electrostriction force to generate pumping to increase the heat pipe capillary limit, but did not provide any experimental evidence. Melcher [121] discussed the phenomena that dielectric liquid tended to fill the region where the electric field is the strongest.

Bryan and Yagoobi [122] performed an experimental study on a monogroove EHD heat pipe and obtained 100% increment in the heat transport capacity in response to heat input variations. Yu et al. [123] conducted an experimental investigation to evaluate the potential benefits of EHD forces on the operation of MHPs. In these experiments, electric fields were used to orient and guide the flow of the dielectric liquid within an MHP from the condenser to the evaporator. The MHP array was manufactured from a 1-mm-thick glass slide. It was 28 mm long and was composed of seven parallel grooves that terminated in plenums on both ends. The grooves and plenums were machined by an ultrasonic milling process. The cross section of each channel was 1 mm wide by 0.6 mm deep, and the spacing between the channels was 1 mm. The experiments indicated that the heat transport capability of the EHD MHPs is increased by up to six times of that of conventional ones. An analytical model was developed to predict the maximum heat transport capability for various electric field intensities, and MHP geometries. There was good agreement between the results of the analytical model and the experimental results for the geometry studied experimentally. The maximum heat transport capacity increased with an increase in the electric field intensities because at high electric field the EHD force became dominant. The model showed that large pore sizes are optimum for more heat transport capacity. Finally, a critical assessment of the experimental results suggested an alternative design capable of achieving as much as a 180 times improvement in the heat transport capacity as compared to traditional MHPs. Later, Yu *et al.* [124] suggested an alternative design capable of achieving as much as a 240 times improvement in the heat transport capacity in comparison to traditional micro heat pipes. They have prepared semimathematical model and validated it with the experimental studies [125].

Suman [126] presented a model for the fluid flow and heat transfer in an EHD-augmented MHP utilizing a macroscopic approach and considering Coulomb and dielectrophoretic forces. The coupled nonlinear governing equations for fluid flow, heat and mass transfer were developed based on the first principles and are solved numerically. The contributions of Coulomb and dielectrophoretic forces studied together, and separately. The analytical expressions for critical heat input and dry-out length obtained. It was observed that with an increase in the electric field intensity, the critical heat input increased. The critical heat input can be increased by 100 times using an electric field. The dry-out length increased with an increase in heat input, viscosity, and friction factor. The effect of Coulomb force was found stronger than dielectrophoretic force. The critical heat input and dry-out length values obtained using the expression developed compared with the experimental results are available in Yu *et al.* [123,124].

D. STEADY-STATE SENSITIVITY ANALYSIS

Suman and Hoda [31] presented the sensitivity analysis of a typical steady-state model of an MHP by using the modified model for a V-shaped MHP. It was found that an increase in the apex angle, inclination, viscosity, sharpness of the corner, and length of the heat pipe reduced the performance of a heat pipe. On the other hand, an increase in the surface tension and contact angle improved the performance. The effect of perturbations in contact angle, radius of rounded ungrooved substrate, apex angle, and inclination were found to be strong. To operate an MHP, a coolant liquid with higher surface tension, lower viscosity, and higher latent heat capacity with few degrees of contact angle should be preferred. The sharpness of the groove was found to be very critical. If the groove was not sharp enough, that is, the radius of ungrooved substrate was more; the MHP might cease to work even before it reached its other operating limits.

IV. Transient Operation of Micro Heat Pipes

A. TRANSIENT MODELS

The study of transient operation in an MHP is important for the startup and shutdown applications, and also to understand the capillary behavior of an MHP. The transient behavior of MHPs under various operating conditions has been under study for quite some time. Next we present them in brief.

Wu and Peterson [19] developed a transient numerical model to predict the thermal behavior of MHPs during start up and to predict the variation in the evaporator thermal load. The model was used to identify, evaluate, and better understand the phenomenon, which governs the transient behavior of MHPs as a function of the physical shape, working fluid, and principle dimensions. Parameters that affect the axial heat transport capacity were evaluated. The model utilized the relationship presented by Collier [90] and later by Colwell and Chang [91] to determine the free molecular flow mass flux of evaporation. The evaporation–condensation rate was assumed to be proportional to the liquid–vapor interfacial area in each section of the MHP. Comparison with steady-state experimental results accurately predicted the steady-state dry-out limit for two different heat pipes. The results indicated that the reverse liquid flow occurred in the liquid arteries during start up and/or rapid transition. A wetting angle was found to be an important factor contributing to the transport capacity. Wu et al. [127] verified the accuracy of the model by Wu and Peterson [19]. The results showed that the numerical model predicted the steady-state behavior but underestimated the transient response. The liquid area as a function of time and position is given in Fig. 12. It is seen that with time, the liquid area in the evaporative section decreases and that increases in the evaporative section.

Sobhan et al. [128] presented a model for the vapor and liquid flow in an MHP with triangular channels having evaporator and condenser sections and water as the working fluid. The governing equations were derived by considering the variation of a cross sectional area of vapor and liquid and incorporating a phase change during the process. Numerical solutions for the transient and steady-state operation of an MHP consisting of an evaporator and a condenser section were evaluated. The energy equation along with other governing equations was solved to yield transient and steady-state distributions of temperature, pressure, and velocity. The effect of the heat input at the evaporator section and the convective heat transfer coefficient at the condenser section was investigated. The effective thermal conductivity of the MHP was also calculated. Similarities and differences of the results compared with those from the literature were discussed. It was shown that there was a significant drop in the vapor temperature at the junction of

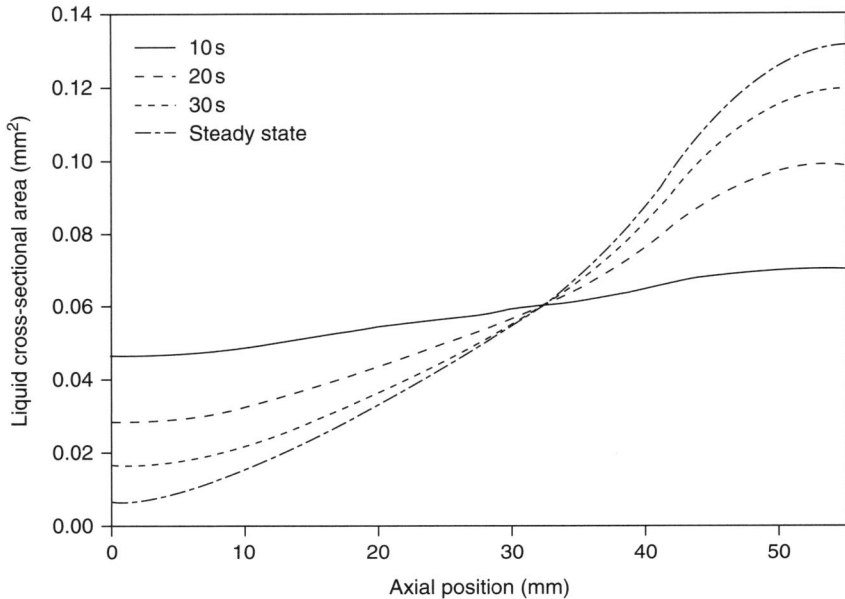

FIG. 12. Liquid cross-sectional area (mm^2) with position (mm) at different time (seconds) from Wu and Peterson [18].

evaporator and condenser sections. The effective thermal conductivity, which compared the overall performance of the MHP, increased under a transient operation. The effective thermal conductivity increased with the power throughput when a constant heat transfer coefficient was assumed for the condenser.

Suman et al. [30] presented a 1D semitransient model for the fluid flow and heat transfer of an MHP of any polygonal shape utilizing a macroscopic approach. The coupled nonlinear governing equations for the fluid flow, heat and mass transfer were developed based on first principles and solved numerically. The transient behavior for various parameters of substrate temperature, radius of curvature, and liquid velocity were studied. The effects of the groove dimensions, heat input, and Q profiles on the studied parameters were evaluated. The steady-state profiles for substrate temperature, radius of curvature, and liquid velocity were also generated. The model-predicted steady-state substrate temperature profile was successfully compared with the experimental results from the previous study by Anand et al. [27].

Suman and Hoda [34] presented the transient mathematical model for the heat transfer and fluid flow of a V-shaped MHP utilizing a macroscopic approach. The coupled nonlinear partial differential equations governing the fluid flow, heat and mass transfer were solved using a novel numerical technique. The variations in fluid flow, heat and mass transfer processes were considered with time. The sensible heat used by the substrate was taken into account using a pseudo lump capacity model. The time required to reach to steady state was found to be dependent on the heat input, and was independent of MHP inclination, groove angle and Q_{ss} profile. Various parametric studies were performed to understand the unsteady state operation of an MHP. The time required to reach dry out was also presented. The model predicted steady-state substrate temperature profile was successfully compared with the experimental results from the previous study by Anand et al. [27].

1. A Transient Model

A typical transient model for an MHP is given below [30,34]. A V-shaped triangular MHP is considered. The schematic of the MHP is shown in Fig. 2. Sufficient amount of the coolant liquid is charged to the system so that at steady state, the grooves at the bottom of the MHP, that is, the cold end remains filled with the coolant liquid such that the radius of curvature can be calculated from the geometry of the system ($R = R_o$). The model gives the transient fluid flow, heat transfer, and relates them to the capillary forces present in the system. The model equations are derived under the assumptions used in the derivation of a steady-state model (Section III.A.1).

Heat supplied by the heater is initially conducted through the solid substrate. The supplied heat to the substrate is later taken up by the coolant liquid in the evaporative section and is transferred by a two-phase heat transfer mechanism, that is, evaporation. Similarly, the vapor is condensed in the condensing section, and the released heat is being taken up by the solid substrate. No interaction of heat between the solid substrate and the coolant liquid has been considered in the adiabatic section. The heat taken up and released by the coolant liquid is approximated using a lump capacity model [129]. The part of heat that is not used by the substrate has been considered as the sensible heat and the convective losses. Hence, the heat flux taken up or released by the coolant liquid is given by

$$Q = Q_{ss}\left(1 - e^{\frac{-t}{\tau}}\right) \tag{22}$$

where Q_{ss} is the input heat flux supplied or taken up by the coolant at steady state, Q is the heat flux used or released by the coolant liquid to the substrate at time t, and τ is the time constant and is defined as $\tau = (\rho_s C_{ps} L^2)/K_s$, which is a characteristic time scale for the heat transfer in the solid substrate.

One corner of a section of a V-shaped MHP of length Δx is taken as the control volume (Fig. 3). The liquid pressure is a function of radius of curvature and the relationship is given by the Young–Laplace equation in the differential form as

$$\frac{\partial P_l}{\partial x} = \frac{\sigma_l}{R^2}\frac{\partial R}{\partial x} \qquad (23)$$

The unsteady state momentum balance in the differential form is given as

$$\rho_l A_1 V_1 \frac{\partial V_1}{\partial x} + A_1 \frac{\partial P_l}{\partial x} + 2L_h \tau_w + \tau_{1,i} R_m - \rho_l g \sin\beta A_1 - \frac{A' A_1}{2\pi h^4}\frac{dh}{dx} + \frac{\partial(\rho_l A_1 V_1)}{\partial t} = 0 \qquad (24)$$

At unsteady state mass balance for liquid in the differential form is given as

$$\frac{\partial(\rho_l A_1)}{\partial t} + \frac{\partial(\rho_l A_1 V_1)}{\partial x} + \frac{Q_v R_m}{\lambda_l} = 0 \qquad (25)$$

The unsteady state energy balance equation in the differential form can be expressed as

$$\rho_l C_{pl} A_1 \frac{\partial T_l}{\partial t} + \rho_l C_{pl} V_1 A_1 \frac{\partial T_l}{\partial x} - QW_b + Q_v R_m = 0 \qquad (26)$$

where Q is the heat flux taken up or released by the coolant liquid from the solid substrate.

The quantity Q is positive when taken up by the coolant liquid (in the evaporative section). It is negative when it is released by the coolant liquid (in the condenser region) to the substrate. It is zero in the adiabatic section as there is no evaporation and condensation in this section. The unsteady state energy balance in the substrate is given as

$$A_{cs} K_s \frac{\partial^2 T_s}{\partial x^2} - QW_b - \rho_s C_{ps} A_{cs} \frac{\partial T_s}{\partial t} = 0 \qquad (27)$$

a. **Nondimensionalization.** For nondimensionalization, the following expressions have been used apart from those used with steady-state model (Section III.A.1), and time constant, $\tau = (\rho_s C_{ps} L^2/K_s)$, T_1^*(dimensionless coolant liquid temperature) $= T_1/T_r$, t^*(dimensionless time) $= t/\tau$. After nondimensionalizing and rearranging, equations can be written as

$$\frac{\partial P_1^*}{\partial X^*} = \frac{1}{R^{*2}} \frac{\partial R^*}{\partial X^*} \tag{28}$$

$$\frac{\partial R^*}{\partial t^*} = -\left[\frac{\tau R^* Q_v R_m L}{2\rho_1 A_1 \lambda_1} + \frac{2\tau V_R V_1^*}{LR^*} \frac{\partial R^*}{\partial X^*} + \frac{\tau R^* V_R}{2L} \frac{\partial V_1^*}{\partial X^*}\right] \tag{29}$$

$$\frac{\partial V_1^*}{\partial t^*} = \frac{\tau g \sin(\beta)}{V_R} + \frac{\tau Q_v R_m V_1^*}{A_1 \rho_1 \lambda_1} - \frac{\tau B_2 V_1^*}{\rho_1 (R_o R^*)^2} - \frac{\tau \tau_{1,i} R_m}{A_1 \lambda_1}$$
$$+ \left[\frac{2\tau V_R V_1^{*2}}{LR^*} - \frac{\tau \sigma_1}{\rho_1 V_R R_o LR^{*2}} + \frac{\tau A'}{2\pi \rho_1 V_R L h_o^3 R^*}\right] \frac{\partial R^*}{\partial X^*} \tag{30}$$

$$\frac{\partial T_s^*}{\delta t^*} = \frac{\partial^2 T_s^*}{\partial X^{*2}} - \frac{Q w_b L^2}{T_R A_{cs} K_s} \tag{31}$$

$$Q_v = \frac{1}{R_m}\left[Q W_b - \frac{\rho_1 C_{pl} V_R V_1^* A_1 T_R}{L} \frac{\partial T_1^*}{\partial X^*} - \frac{\rho_1 C_{pl} A_1 T_R}{\tau} \frac{\partial T_1^*}{\partial t^*}\right] \tag{32}$$

$$Q = Q_{ss}(1 - e^{-t^*}) \tag{33}$$

The steady-state heat flux is given as

$$Q_{ss} = \frac{Q'}{fW_b L} \tag{34}$$

Here geometrical parameters A_1, W_b, R_o, and B_2 are presented in Eqs. (17)–(21).

The dimensionless boundary can be written as

at the cold end: $(X^* = 1)$, $R^* = 1$, $P_1^* = (P_{vo}/P_R) - 1$, $T_s^* = T_{con}/T_R$, $V_l^* = 0 \; \forall \; t^*$.

at the hot end: $(X^* = 0)$, $Q_{heater} = -\frac{K_s A_{cs} T_R}{L} \frac{\partial T_s^*}{\partial X^*}|_{X^*=0}, \; \forall t^*$.

Initial conditions:

at $t^* = 0$, $T_s^* = T_{con}/T_R$, $V_l^* = 0$, $P_1^* = \frac{P_{vo}}{P_R} - 1 - \frac{\rho_1 g L(1 - X^*)\sin(\beta)}{P_R}$,

$R^* = \frac{P_R}{(P_{vo} - P_R P_1^*)}, \; \forall X^*$.

b. Results and Discussion. To present the transient analysis of a V-shaped MHP, the same system that is used for the steady state has been taken.

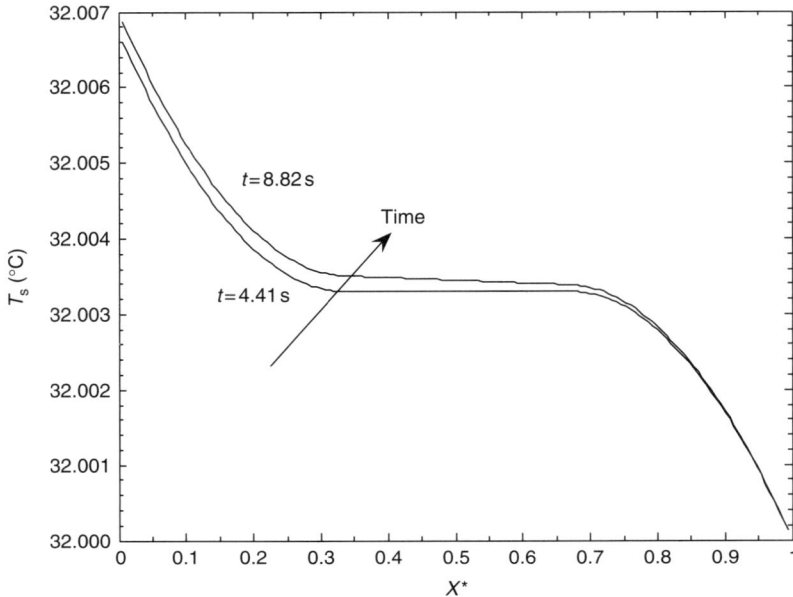

FIG. 13. Temperature profiles of the substrate as a function of dimensionless axial position at different times.

Temperature profiles of the substrate as a function of dimensionless axial position at different times are presented in Fig. 13. Initially, the substrate temperature of the heat pipe is constant. As the heater starts giving heat to the substrate, the substrate temperature linearly starts to increase from the cold end to the hot end (refer to the profile at $t = 0.01$ s). This is because initially no heat is taken by the coolant liquid. As the heat taken up by the coolant liquid increases, the chair-shaped temperature profile is established which decreases the adiabatic section temperature (refer to the profiles at $t = 4.41$, 8.82, and 13.22 s). At steady state, the temperature difference between the hot and the cold regions is small. The adiabatic section temperature is also small. Therefore, it is said that an MHP can remove heat without generating a higher temperature difference and this property of an MHP makes it very promising for many potential applications. It is observed that initially the substrate temperature is increasing with time and then it is decreasing. As explained above, initially the heat is being supplied by the heater to the substrate, and the heat is not taken up by the coolant liquid. Therefore, it is increasing. After sometime, the heat taken up by the coolant liquid dominates and the temperature profile starts decreasing. Finally, the steady-state profile has been established.

The profiles of the dimensionless radius of curvature along the dimensionless axial position of the heat pipe (R^* vs. X^*) at different times are presented in

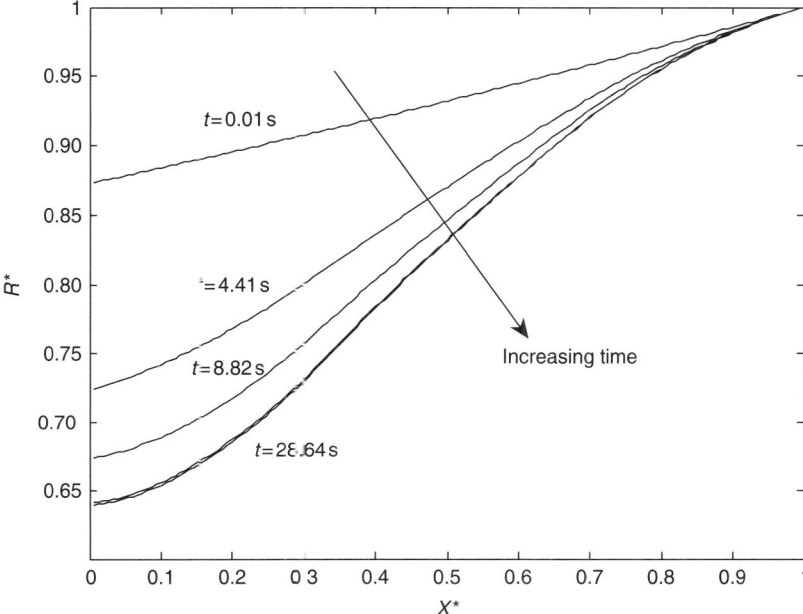

Fig. 14. Dimensionless radius of curvature along the dimensionless axial position at different times.

Fig. 14. Figure 14 shows that initially the radius of curvature is decreasing from the cold to the hot end to balance the pressure difference generated by gravity. Once the evaporation of the coolant liquid starts, the radius of curvature is decreasing with time to generate enough driving force to pull the liquid from the cold to the hot end. Initially, the heat taken by the coolant liquid is less and therefore less driving force is required. Thus, the difference in curvature is less (refer to the profile at $t = 0.01$ and 4.41 s). Once the heat input taken by the coolant liquid is almost constant, the amount of the coolant liquid to be evaporated becomes constant and this requires a fixed driving force. Therefore at steady state, the radius of curvature is monotonically decreasing with position and it remains constant with time (refer to the profile at $t = 28.64$ s). At steady state, the value of R^* is increasing from the hot end to the cold end to provide sufficient capillary pumping for the fluid flow in the MHP.

The radius of curvature is decreasing as heat supplied to the coolant liquid increases with time. Therefore, the difference between the vapor and the liquid pressure should increase. It means that the liquid pressure should decrease with time since the vapor pressure is considered constant. The variation of the liquid

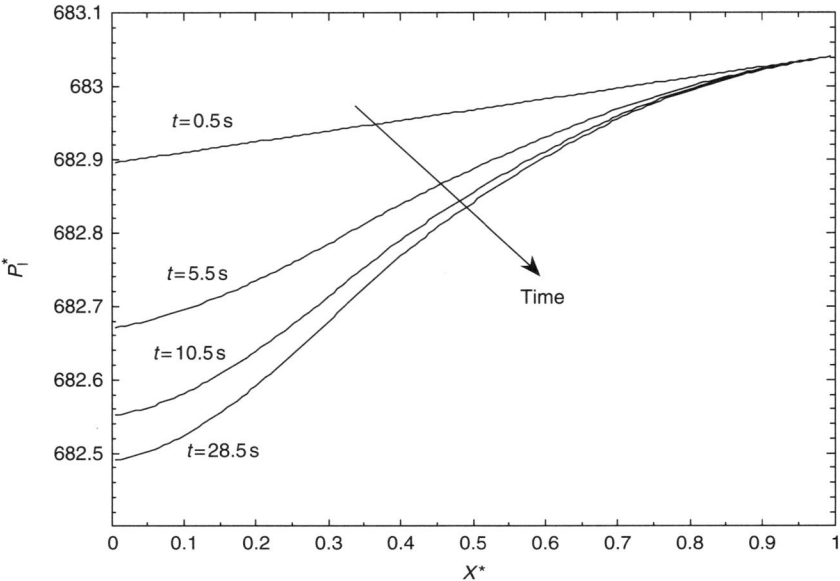

FIG. 15. Variation of liquid pressure with dimensionless axial position at different times.

pressure with the dimensionless axial position at different times is presented in Fig. 15 and consistent result has been obtained. Initially, a lower pressure difference is required to sustain the flow (refer to the profile at $t = 0.01$ s). With an increase in time, the heat supplied to the coolant liquid increases and to sustain evaporation, more liquid at the evaporative section is required which demands a larger pressure drop between the cold end and the hot end to sustain the increase in coolant liquid flow. Therefore, the pressure difference between the hot and the cold end increases with time (refer to the profile at $t = 0.01$ and 4.41 s).

The variation of the axial liquid velocity with dimensionless axial position at different times is presented in Fig. 16. The direction of the liquid velocity is from the cold to the hot end. Initially, there is no heat input to the coolant liquid, and therefore, the liquid velocity is zero. As the input heat to the coolant liquid is increasing, the liquid starts flowing from the cold end to the hot end due to capillary pumping. With time, the amount of heat to be transferred is increasing and so is the liquid velocity (refer to the profile at $t = 0.01$ and 4.41 s). After sometime, the amount of liquid to be evaporated becomes constant. This requires a constant pressure drop, and hence the radius of curvature reaches steady state. This results into the liquid velocity to reach steady state. The axial liquid velocity is always zero at the hot end,

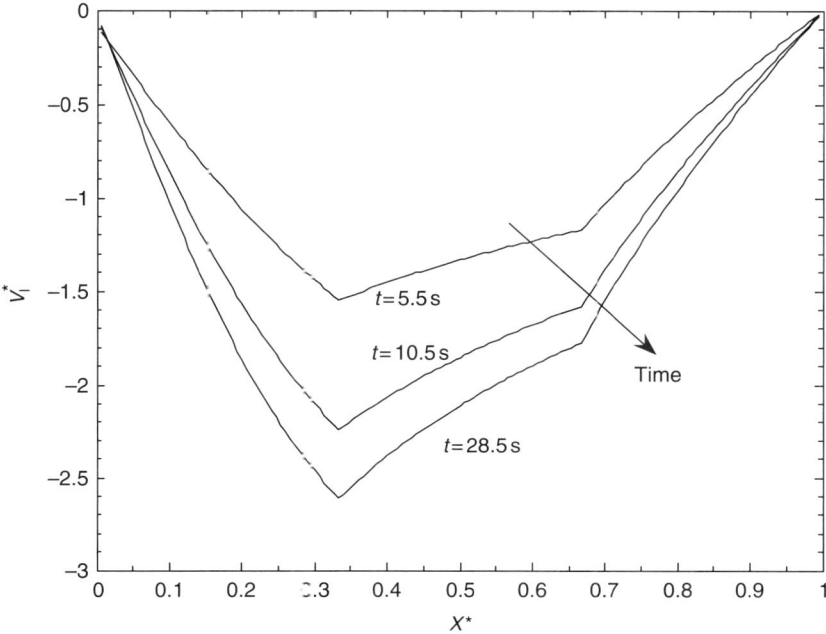

FIG. 16. Variation of axial liquid velocity with dimensionless axial position at different times.

and increases in the evaporative region. This is due to the cumulative effect of replenishing the amount evaporated, throughout the evaporative section. In the adiabatic section, although there is no evaporation and condensation, the liquid velocity decreases slightly. This is because of a slight increase in the radius of curvature in the adiabatic section (hence an increase in the liquid flow area) and the friction loss. In the condenser section, there is a further decrease of the liquid velocity because the condensation results in a sharper increase in the radius of curvature and so does the liquid flow area.

B. Transient Experimental Study

Wu et al. [127] conducted an experimental investigation on several small, tapered MHPs specifically designed for the use in a thermal control of ceramic chip carriers to verify the operation, measure the performance limits, and transient behavior. Several MHPs were evaluated for start up or rapid changes in the thermal load. The experimental data was compared with the results from a previously developed analytical model to determine the accuracy of the model and verify the predicted results. The transient

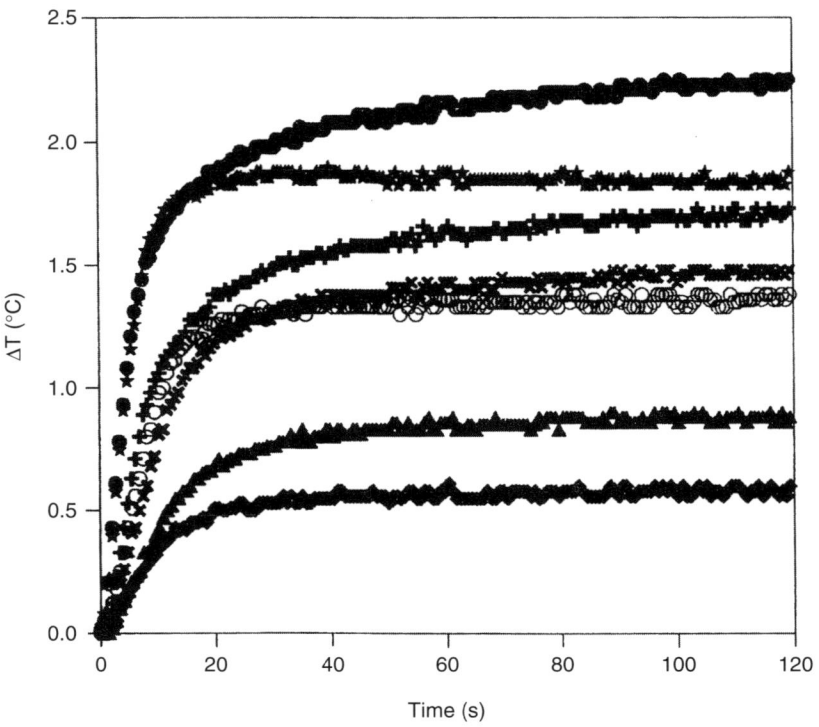

- ●— Thermocouple 1
- —+— Thermocouple 2
- - ★ - Thermocouple 3
- —○- - Thermocouple 4
- —✖— Thermocouple 5
- - -▲ - Thermocouple 6
- —◆— Thermocouple 7

FIG. 17. Transient experimental distribution (power = 0.058 W, $T_c = 23°C$) as taken from Wu *et al.* [127].

temperature profiles of the MHP for an input power of 0.508 W and a condenser temperature of 23°C are shown in Fig. 17. It can be concluded from the figure that the heat pipe reaches steady state very quickly with approximately 90% of the temperature being reached in <20 s.

Ma and Peterson [24] conducted an experimental investigation to measure the capillary heat transport limit in small triangular grooves similar to those used in MHPs. One end of a plate was immersed in a liquid pool and held stationary at a specified tilt angle. The liquid was wicked up the grooves by the capillary pressure against the gravitational force and the frictional pressure drop resulting from the continuous flow of liquid. An axial adiabatic region was included between the point of origin and the junction point to simulate a micro heat pipe. Experimental measurements were made to determine the heat transport and the unit effective area heat transport in triangular grooves with an apex angle of 60°. They showed that the heat transport decreased with an increase in the tilt angle and the effective length. It was also shown that increasing the dimensions of an individual groove could not improve the overall heat transport due to the decrease in the unit effective area heat transport. The experimental results indicated that the uniform heat flux assumption made previously may not be entirely correct and the assumption of constant radius of curvature was applicable when the Bond number was much smaller than 1.

C. TRANSIENT SENSITIVITY ANALYSIS

Suman [33] presented a sensitivity analysis of the transient model of an MHP. The effect of gravity force on the transient behavior of MHP was studied by varying acceleration due to gravity and inclination. The radius of curvature with time monotonically decreased with time. It was seen that with higher opposing gravity force, the time to reach steady state increased and the performance of the MHP deteriorated (as variation in R^* between the hot end to the cold end is more). This was because that a part of the capillary pumping overcomes the gravity and the heat pipe needs less amount of coolant liquid since the initial radius of curvature was less. With an increase in surface tension, the time to reach steady state decreased and the performance factor increased. As with an increase in surface tension, the capillary pumping available for flow increased. With an increase in viscosity, the time to reach steady state increased and the performance factor decreased. This was due to an increase in viscosity, which increases the friction loss. With an increase in contact angle, the time to reach steady state decreased and the performance factor increased. This was because that with an increase in contact angle, the accumulation of liquid in the groove for heat transfers increased.

With an increase in length of an MHP, the time to reach steady state increased and the performance factor decreased. This was because that with an increase in the length of an MHP the frictional loss increased. With an increase in frictional factor, the time to reach steady state increased and the performance factor decreased. This was because that with an increase in friction factor the frictional loss increased. With an increase in the length of the adiabatic section, the time to reach steady state increased. This was because that with an increase in the length of the adiabatic section the frictional loss increased. With an increase in fraction of ungrooved area, the time to reach steady state increased and the performance factor decreased. This was because that the more the ungrooved area; the higher the fluid velocity and hence, the more the frictional loss.

Suman [33] also performed the simulations for a sinusoidal time-varying perturbation in viscosity, surface tension, and friction factor. The disturbances were obtained in the transient profiles, but it was not decaying or growing with time rather, it was steady with time. The effect of the time-varying disturbances was similar to the effect discussed with the time invariant variations in these variables. For example, when the surface tension was increased, the performance improved and the opposite was true for the viscosity.

V. Operating and Design Parameters

A. Fill Charge

The amount of the coolant liquid to be charged in an MHP, known as the fill change, has a significant effect on its performance [130]. The charge optimization was studied in Refs. [131,132].

The equation for the fill charge calculation was given by Wu and Peterson [18]. The fill charge (m_f) can be calculated from the following equation:

$$m_f = \rho_l L \int_0^1 A_l \, dX^* + \rho_g L \int_0^1 A_g \, dX^* \\ = (\rho_l - \rho_g) L B_1 R_o^2 \int_0^1 (R^*)^2 \, dX^* + \rho_g A_{cs} L \tag{35}$$

While expending the above equation into the three contributions to the fill charge,

$$m_f = \underbrace{\rho_l L \int_0^{L_f^*} A_l \, dX^* + \rho_g L \int_0^{L_f^*} A_g \, dX^*}_{\text{Flooded region}} + \underbrace{\rho_l L \int_{L_f^*}^{1-L_d^*} A_l \, dX^* + \rho_g L \int_{L_f^*}^{1-L_d^*} A_g \, dX^*}_{\text{Operating region}}$$

$$+ \underbrace{\rho_g L \int_{1-L_d^*}^{1} A_g \, dX^*}_{\text{Dry region}}$$

(36)

Further simplification leads to the following equation:

$$m_f = \rho_l L B_1 R_o^2 \int_0^{L_f^*} dX^* + \rho_g L (A_{cs} - B_1 R_o^2) \int_0^{L_f^*} dX^* + \rho_l L B_1 R_o^2 \int_{L_f^*}^{1-L_d^*} R^{*2} \, dX^*$$

$$+ \rho_g L \int_{L_f^*}^{1-L_d^*} (A_{cs} - B_1 R_o^2 \bar{R}^{*2}) dX^* + \rho_g L \int_{1-L_d^*}^{1} A_{cs} \, dX^*$$

(37)

Here, B_1 was a geometrical parameter and it was equal to A_1/R^2. From the above expression (Eq. (36)), we saw that the fill charge had three contributions: first from the flooded region, second from the operating region and third from the dry-out length region. It was to be noted that the contribution from the flooded region and the dry-out region deteriorated in the performance of the MHP. We always aimed to minimize the flooded and the dry-out length. If the coolant liquid was overcharged, a portion of the condensing section was flooded and it increased the liquid pool thickness. An increase in the liquid pool thickness increased the heat transfer resistance of the coolant liquid. Flooding in a portion of condenser lead to a smaller condenser section, and therefore a higher rate of heat removal from the condenser region was required. If the coolant liquid was undercharged, the area available for the flow of the liquid was less. This increased the fluid velocity resulting into a higher frictional loss. Additionally, the coolant liquid available for the evaporation was less, and therefore the onset of the dry out was expected to occur at a lower heat input. Its detailed study was presented in Suman [33].

1. Optimal Fill Charge

The optimal fill charge is the amount of coolant liquid required to meet the boundary conditions, $R^* = 1$ at $X^* = 0$, at a particular heat input without generating any hot spot (i.e., $R^* = 0$ at $X^* = 1$). In other words, the length of the flooded and the dry regions is zero, and the cold end is completely filled, that is, $R^* = 1$ at $X^* = 0$ and $R^* = 0$ at $X^* = 0$. The expression for the optimal fill charge is given as follows [33]:

$$m_f = \underbrace{\rho_l L \int_0^1 A_l \, dX^* + \rho_g L \int_0^1 A_g \, dX^*}_{\text{Operating region}} \quad (38)$$

$$= \rho_l L B_1 R_o^2 \int_0^1 R^{*2} \, dX^* + \rho_g L \int_0^1 (A_{cs} - B_1 R_o^2 R^{*2}) \, dX^*$$

The optimal fill charge was calculated by using numerical integration to solve Eq. (38). The dimensionless radius of curvature (R^*) as a function of X^*, which satisfied $R^* = 1$ at $X^* = 0$ at a particular heat input, could be calculated by solving dimensionless Eqs. (28)–(34). Equation (38) along with the dimensionless model equations (Eqs. (28)–(34)) suggested that the optimal charge increased with an increase in heat input, contact angle, and length of heat pipe, and decreased with an increase in inclination, acceleration due to gravity, and surface tension.

2. Charging Scheme

Lee et al. [133] presented a method of charging. The characterization of an MHP system, integrated with a local heater and temperature and capacitive microsensors, was presented. Two liquid charging schemes based on a single hole, requiring vacuum environment, and a pair of holes, utilizing capillary forces were compared. Taking advantage of the great disparity between the dielectric constants of liquids and gases, capacitance sensors were used for void fraction measurements.

It was difficult to control the phase content of a liquid–gas mixture in an MHP, and therefore a calibration technique based on a traveling water–air interface due to evaporation was introduced. The integrated sensor capacitance for pure water was found to depend on measurement frequency, temperature, and ion concentration, exhibiting trends that were different from previous reports. The measured temperature and void fraction distribution along the MHPs were consistent with the two-phase flow patterns recorded during the microsystem operation.

B. Liquid Height

The coolant liquid height (h) is defined as the maximum thickness of the coolant liquid pool. The expression for the coolant liquid height is given as [33]

$$h = R_o R^* \left[\frac{\cos(\alpha + \gamma)}{\tan \alpha} + \sin(\alpha + \gamma) - 1 \right] \tag{39}$$

It was seen from Eq. (39) that the coolant liquid height depended on the contact angle of the coolant liquid–substrate system, the radius of curvature, and the groove geometry. The coolant liquid height should be considered while designing an MHP.

C. Coolant Liquid Temperature

The liquid temperature was calculated in Suman [33] based on a predefined rate of evaporation, condensation, and liquid thermal resistance. The temperature difference between the substrate and the coolant liquid depended on the thermal resistance of the coolant liquid. The temperature of the midpoint of the coolant liquid was calculated and it can be extrapolated/interpolated to get the coolant liquid temperature at any location in the liquid pool. The coolant liquid temperature at the midpoint is given below:

$$T_1^* = T_s^* - \frac{Q_{avg} R_{th}}{T_R} \tag{40}$$

where $Q_{avg} = (QW_b + Q_v L_h)/2L_h$ and $R_{th} = (A_1/2L_h)/K_1$.

D. Effect of Surface Tension Gradient on a Micro Heat Pipe

Suman [134] studied the effect of surface tension gradient on the performance of an MHP. The force due to a surface tension gradient was considered in the momentum balance equation. The analytical expressions for the radius of curvature, liquid pressure, liquid velocity, and critical heat input were obtained. The favorable and the unfavorable effects of the surface tension gradient were shown on the critical heat input, liquid pressure, and radius of curvature. The effect of the surface tension gradient was shown to be increased with an increase in the number of sides of a polygonal heat pipe. For the limiting case, the ratio of the surface tension at the hot and the cold end was obtained when the opposing surface tension gradient flow stopped the fluid flow due to the capillary pumping. The adverse effect of the surface tension gradient could be due to the liquid temperature variations. The favorable situations could be prepared selecting suitable surfactant, additives, and surface charge etc., where the surface tension gradient can facilitate the liquid flow in an MHP and then it would improve the performance of an MHP.

E. FRICTION FACTOR AND SURFACE ROUGHNESS

It was known that the surface roughness did not affect the laminar flow in macrochannels. However, as the channel size decreased to the order of few microns, the effect of surface roughness microchannels became important. For the same roughness, the influence was stronger in a smaller size channel. Nusselt number and the friction factor of a high roughness microchannel increased faster than that of a low roughness microchannel with increasing Reynolds number. This was because the disturbance in the boundary sublayer by the roughness, which was more significant at high Reynolds numbers. High liquid–solid contact in microfluidics made correct evaluation of friction factor a potential research field. As expected, it was no longer identical to channel or any macro flow. The anisotropic echants frequently leave rough surfaces in the microchannels. Chen and Cheng [135] presented a detailed discussion on the friction factor in microchannels. White [136] presented a friction factor for laminar flow and defined the friction factor such that fN_{Re} = constant and the constant was called Poiseuille number with f being the friction factor and N_{Re} the Reynolds number. The analytical solution was given by Ma and Peterson [137]. The measurement of roughness was defined in [138]. The difference in the surface roughness profile for different temperatures and concentrations were given in Kang *et al.* [139].

Ma and Peterson [140] obtained an analytical solution for the friction factor of a laminar incompressible flow in microchannels of irregular cross section with two arbitrary boundaries. Their study confirms that the shape of the cross-sectional area had a great influence on the friction constant for the laminar flow of water inside the silicon microchannels. In the study of Ma and Peterson [140], using a combination of analytical techniques and numerical integration methods, an expression governing laminar, fully developed fluid flow occurring in very small ducts of irregular cross section was developed and solved. By changing the functions that describe the boundaries, the friction factor and Reynolds number product was readily obtained for a number of different types of cross sections. Using this method, the friction factor and Reynolds number product occurring in an MHP where the cross section varied from the evaporator to the condenser was solved without the need to reformulate the problem to determine the analytical solution, or reestablished the grid mesh to find numerical results for a given location along the axial direction.

Wu and Cheng [100] developed a correlation based on a few hundred experiments that relates laminar friction constant of the trapezoidal microchannels in terms of the cross-sectional aspect ratio (W_b/W_t) given by

$$fN_{Re} = 11.43 + \exp\left(2.67\frac{W_b}{W_t}\right) \tag{41}$$

The quantity (W_b/W_t) = 0.0 corresponded to a triangular channel and (W_b/W_t) = 1.0 corresponded to a highly flat trapezoidal microchannel.

The friction coefficient for the 2D laminar flow of channel with axial trapezoidal grooves was also given by Suh et al. [141]. An approximate relation that was previously utilized for the friction in the liquid was modified to obtain accurate agreement with the results for trapezoidal and sinusoidal grooves. The results for the friction in the trapezoidal grooves could be summarized as

$$(f\text{Re})_v = \left(-0.94 + 3.8e^{\frac{\pi h_c}{2h}} + \frac{11.8}{1 + \sin\alpha}\right) \\ + \left(\frac{W}{L}\right)^2 \left(52 + 4.6e^{\frac{\pi h_c}{2h}} + \frac{0.89}{1 + \sin\alpha}\right) \tag{42}$$

and for the sinusoidal grooves as

$$(f\text{Re})_v = \left(22.2 + 2.53e^{\frac{\pi h_c}{2h}} - \frac{4.1}{1 + \sin\alpha}\right) \\ + \left(\frac{W}{L}\right)^2 \left(-29.7 + 5.43e^{\frac{\pi h_c}{2h}} + \frac{42.3}{1 + \sin\alpha}\right) \tag{43}$$

F. Fabrication of Micro Heat Pipes

Fabrication of MHPs was a key issue for its development and applicability. The fabrication methods for MHPs were very important for accurate design as the length of the MHP was on the order of magnitude of 30–100 μm. Investigations were done which have focused on the fabrication, development, and testing of the MHPs. Several fabrication techniques were proposed by which MHPs could be incorporated as an integral part of silicon or gallium arsenide wafers. Initially, a V groove was investigated and later it was found that even inside a polygon, all corners would behave identically irrespective of their orientation since force generated due to surface tension at the liquid vapor interface was much higher than gravity force [30].

Gerner [142] proposed the formation of a series of parallel triangular grooves using an etching process. In the etching process, an orientation-dependent etching (ODE) was used to produce a series of parallel microchannels in a silicon wafer. The ODE process allowed etching of silicon wafers in one particular direction at a higher rate as compared to the other directions. This was due to the fact that the crystal structure of silicon was in the diamond cubic group which caused the crystallographic directions to be perpendicular to the crystal planes. As a result, the packing density was considerably higher in one direction as compared to the other directions and an etching rate was faster in that direction.

Mallik et al. [143] etched a series of V-shaped microchannels in silicon using a solution of KOH–normal propanol–water. The V-shaped channels had a depth of 80 µm and a width of approximately 84 µm. After etching the microchannels, a 200-µm-thick clear glass cover plate was bonded to the surface to form the closed triangular channels using either an ionic bonding process or an ultraviolet bonding process.

Mallik et al. [144] fabricated a vapor-deposited micro heat pipe (VDMHP) array, which was an integral part of semiconductor devices to act as efficient heat spreaders by reducing the thermal path between the heat sources and heat sink. They accomplished the fabrication of the VDMHP by first establishing a series of grooves in a silicon wafer. ODE using a KOH–1-propanol–H_2O solution on a (100) wafer with a (111) flat covered with an oxide mask, resulted in grooves 25 µm wide and 25 µm deep with sharp, perpendicular edges. The wafers were predeposited with a layer of chromium followed by a layer of gold to improve the adhesion characteristics. Dual-electron beam vapor deposition, followed by planetary process using molybdenum crucibles were used to deposit copper 31.5–33.0 µm thick, and provide complete closure of the grooves. A glass coverslip was bonded on the top of the deposited layer. The grooves were finally charged and sealed. A computer model simulation and modeling of evaporated deposition profiles (SAMPLE) to optimize the metal step coverage and successfully predict the cross-sectional profile of the VDMHP.

Kang et al. [145] proposed radial grooved MHP in which the use of the upper layer of vapor was divided from the bottom layer of liquid by a thin layer. The method of using an interface to separate the vapor phase path and the return path of the working fluid effectively improved the entrainment limits created by shear force at the interface between vapor phase and fluid phase and increased the maximum heat transmission of the MHP. They used 80°C, 40% weight percentage KOH etchant on a magnetic rotor rotating at the speed of 400 rpm to perform etching of top and bottom layers and chips at central portion. An Au–Si eutectic bonding was used with optical fiber alignment technique to form the three pieces of the structure into one body. After completion of etching of the MHP, the three wafers were placed into buffered oxide etch (BOE) solution to remove the oxide layer residue. After rinsing with deionized (DI) water, a solution of 3:1 volume mixture of vitriol and hydrogen peroxide was used to clean the wafer surface. This was followed by rinsing with DI water and blowing with nitrogen to dry the wafer. Experiments were performed to evaluate the performance of wafers with three different wafer fill rates at different input powers.

Lee et al. [133] proposed the design and fabrication of an integrated MHP system consisting of a heater, an array of heat pipes, and temperature and capacitive sensors. They fabricated the MHPs using the etching process and covered the grooves with a nitride membrane using a wafer bonding and etch back technique. The flat nitride surface allowed the integration of electrical or mechanical devices utilizing standard complementary metal oxide semiconductor complementary metal-oxide-semiconductor (CMOS)-compatible micromachining techniques. However, the nitride membrane was not transparent enough to allow clear observation of two-phase flow patterns in the MHPs. Therefore, a glass wafer was selected as an alternative cover to facilitate good visualizations of the evolving flow patterns inside the pipes.

Weichold et al. [58] proposed another fabrication technique which utilized a dual source vapor deposition process. A series of square or rectangular grooves were either machined or etched in a silicon wafer. The grooves were closed using a dual E-Beam vapor deposition process to create an array of long and narrow channels of triangular cross sections and open on both the ends. The MHPs were lined with a thin layer of copper to reduce problems associated with the migration of the working fluid to the semiconductor material. The vapor deposition process was optimized by using a model and an array of VDMHPs was constructed successfully in grooves with an aspect ratio of 1.0.

Berre et al. [146] used anisotropic chemical etching process and direct silicon wafer-bonding technique for the fabrication of an MHP. The fabrication began with thermal growth of a 1.5-μm oxide layer on the device wafer. The oxide on both sides of the device wafer was patterned in order to be used as an etching mask and V grooves were etched using a 40 wt% aqueous KOH solution at 60°C. Anisotropic etching proceeds until it stopped on the planes. After stripping the oxide, the patterned wafers were bonded. Plain silicon wafers were used to seal the heat pipe array.

Moon et al. [147] manufactured an MHP with a polygonal cross section via drawing. The MHP had flat or concave sides to allow working fluid to flow by capillary force generated at the edges of the MHP. Another MHP was manufactured by forming a plurality of through holes with a polygonal cross section in a metal plate via extrusion, in which each of the through holes has flat or concave sides to allow working fluid to flow by capillary force generated at the edges of each of the through holes. The MHPs can be manufactured easily via drawing or extrusion, can induce strong capillary force through simple structural modifications, and provide superior cooling effects.

Lee et al. [148,149] presented the design and fabrication of an integrated MHP system consisting of a heater, an array of heat pipes, and temperature arid capacitive sensors. Taking advantage of the large difference between the dielectric constants of liquid and vapor, the integrated capacitor can be used for void-fraction measurements in two-phase flows. Both CMOS-compatible and glass-based fabrication technologies are reported. In the CMOS-compatible technology, the heat pipes were capped by a thin nitride layer using wafer bonding and etch back technique. In the glass-based technology, the heat pipes were covered by a glass substrate using die-by-die anodic bonding to allow visualization of the two-phase flow patterns. This approach also resulted in a significant reduction of the parasitic capacitance, thus enhancing the sensitivity of the capacitance sensor.

Kang et al. [150] described radial grooved MHPs with a three-layer structure. The MHPs were designed to allow separation of the liquid and vapor flow to reduce the viscous shear force. The $5 \times 5\,\text{cm}^2$ MHP array was fabricated by using bulk micromachining and eutectic bonding techniques on 4-in (1 0 0) silicon wafers. Experiments were undertaken to evaluate the performance of wafers with three different wafer fill rates at different input powers. The heater was glued below the evaporator section, pumped cold water through a square copper heat exchanger above the heat pipe, and pasted 15 K-type thermocouples on both sides of the MHP structure to record the variations of surface temperature. After the evaluation, the MHP with 70% fill rate showed the best performance as compared to samples with smaller fill rates.

G. Performance Factor

The performance factor gives the quantitative measurement of the performance of an MHP. The performance factor, B, was defined in Suman and Kumar [32] as

$$B = \frac{\sigma_1 R_0^3 \rho_1 \lambda_1 B_1}{B_2 L Q_{\text{in}}} \quad (44)$$

This expression captured the property of the coolant liquid using surface tension, viscosity (in B_2), contact angle (in B_2), density, and latent heat of vaporization. Higher the latent heat of vaporization, density, and surface tension, the performance factor increased, while with an increase in viscosity, it decreased. Hence, it was used to study the performance of an MHP when the thermophysical properties of a coolant liquid change. The performance factor also had some geometric constants like B_1, B_2, and L. This measured the effect of the geometric parameters. For example, the performance factor was proportional to B_1, which measured the liquid accommodation capacity of an MHP and hence the performance of an MHP increased with B_2. The higher length, the lower the performance factor. This was because that with an increase in the length of an MHP, the frictional loss increased. Therefore, the performance factor measured the effect of geometric parameters as well.

The higher the value of the performance factor, the better the performance of an MHP, that is, a higher critical heat input. Thus, a combination of thermophysical properties of the coolant liquid, design parameters, and contact angle of the liquid–substrate system that gave a higher value of B should be chosen for a fixed heat input. Therefore, it was suggested that the value of B should be considered while designing an MHP.

VI. Design of Microgrooved Heat Pipes

There are many factors to be considered while designing an MHP. For a working fluid, the compatibility, thermal stability, wettability, vapor pressure, high latent heat, high conductivity, low viscosity, high surface tension, and accepted freezing point have to be considered. A criterion for the selection of a coolant liquid were given in Faghri [12], Dunn and Reay [9], and Heine and Groll [151]. Dunn and Reay [9] defined merit number (Me) for the coolant liquid selection:

$$\mathrm{Me} = \frac{\rho_l \sigma_l \lambda_l}{\mu_l} \tag{45}$$

For a sealed container, the compatibility, thermal conductivity, wettability, strength, porosity, and fabrication have to be considered. In addition to these characteristics, which are primarily concerned with the internal effects, the container material must often be resistant to corrosion (resulting from interaction with the environment) and malleable (to be formed into the appropriate size and shape). Apart from these considerations, operating temperature range, diameter, power limitations, thermal resistances, and operating orientation should also be considered. Sugumar et al. [152] reported that an MHP operated effectively by achieving its maximum possible heat transport capacity only if it was to operate at a specific temperature. In reality, an MHP might be required to operate at temperatures different from that temperature. In the study of Sugumar *et al.* [152], the heat transport capacity of an equilateral triangle MHP was investigated. The MHP was filled optimally with working fluid for a specific design temperature and operated at different operating temperatures. For this purpose, water, pentane, and acetone were selected as the working fluids. The optimal charge level of the MHP was dependent on the operating temperature. Furthermore, the results also showed that if the MHP was to be operated at temperatures other than its design temperature, its heat transport capacity was limited by the occurrence of flooding at the condenser section or dry out at the evaporator section, depending on the operating temperature and type of working fluid. When the MHP was operated at a higher temperature than its design temperature, the heat transport capacity increases, but it was limited by the onset of dry out at the evaporator section. However, the heat transport capacity decreased if it was to be operated at lower temperatures than its design temperature due to the occurrence of flooding at the condenser end. However, the design issues were reduced to two major considerations by limiting the selection to copper/water heat pipes for cooling electronics. These considerations were the amount of power the heat pipe was capable of carrying and its effective thermal resistance.

A conventional heat pipe, which operates on a closed two-phase cycle, consists of a sealed container lined with a wicking material. The container of the heat pipe can be constructed from metals, ceramics, composite materials, or glass. In all applications, careful consideration must be given to the material type, thermophysical properties, and compatibility. For example, the container material must be compatible with the working fluid, strong enough to withstand pressures associated with the saturation temperatures encountered during heat addition and have a high thermal conductivity.

A. LIMITATIONS TO HEAT TRANSPORT

The most important heat pipe design consideration is the amount of power that a heat pipe is capable of transferring. Heat pipes can be designed to carry a few watts or several kilowatts depending upon the application. Heat pipes can transfer much higher power for a given temperature gradient than even the best metallic conductors. If driven beyond its capacity, the effective thermal conductivity of the heat pipe was significantly reduced. Therefore, it is important to ensure that the heat pipe is designed to safely transport the required heat load.

The maximum heat transport capability of a heat pipe is governed by several limiting factors which must be addressed while designing a heat pipe. There are eight heat transport limitations of a heat pipe. These heat transport limits, which are a function of a heat pipe operating temperature,

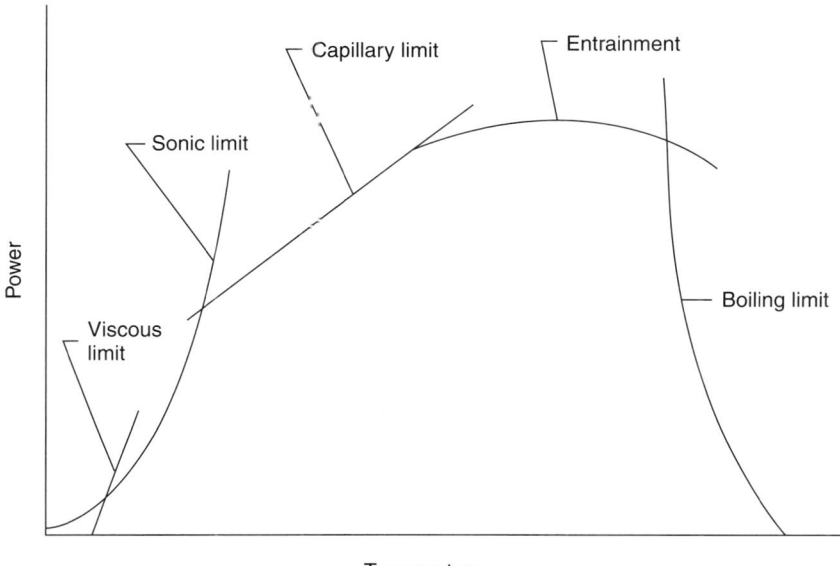

FIG. 18. Typical heat pipe limits.

include viscous, sonic, capillary pumping, entrainment or flooding, boiling, condenser, vapor continuum and frozen start-up. A typical heat pipe limit is presented in Fig. 18. It is possible to ensure that the heat pipe can transport the required thermal load, improve the design and material selection process, and provide a heat pipe that can function within a specific operating temperature range both effectively and reliably. For effective cooling, a heat pipe must operate within its limits [40].

1. Capillary Limit

The capillary action is caused at the evaporator due to the concave meniscus with higher pressure in the vapor than liquid. The higher liquid pressure is communicated to the condenser where liquid and vapor pressures are nearly equal. This drives the liquid from the condenser to the evaporator. The capillary pressure is responsible for the fluid circulation in a heat pipe. It is the limitation of fluid flow for a given capillary pressure difference generated, when this difference is exactly met by the pressure required for the flow of fluid. The heat input corresponding to this is called the critical heat input. Any higher value of heat input generates a dry spot in a heat pipe, and the length of the dry region is called the dry-out length. Therefore, at the critical heat input, we have zero dry-out length and for any higher heat input it increases. For a safe operation of a heat pipe, the capillary pumping head developed should be greater than the required pressure drop for the fluid flow.

The capillary limit for an MHP was investigated. Ma and Peterson [153] presented a mathematical model for predicting the capillary limit of an MHP. In this model, a method for determining the effect of the vapor–liquid frictional interaction on the liquid flow was included. A 2D model for the vapor flow friction factor was developed to obtain the friction factor for the vapor channel with an irregular shape. A control volume technique was employed to determining the flow characteristics of the liquid flow in the MHP. The results indicated that the contributions of the vapor–liquid frictional interaction were very important in determining the maximum heat transport capability, and depend on the operating temperature.

a. Critical Heat Input. The critical heat input for an MHP, where the flow is sustained by capillary pumping, is defined as the heat input when the flow resulting from the curvature change is not able to meet the flow requirement due to higher rates of evaporation. For such a case, the radius of curvature of the liquid meniscus at the hot end reaches a value very close to zero and the device approaches its operating limit. This limit is called the capillary limit and it reaches first in many practical applications. From the Young–Laplace equation, it is clear that for the device to operate properly (no generation of dry spot) the radius of curvature should decrease monotonically from the cold to the hot end (dR^*/dX^* should be always positive according to the coordinate system used herein). The value of R^* at the hot end or the gradient of R^* can also predict the operating limit of a heat pipe, that is, the critical heat input for a system controlled by capillary pumping. There are many methods to calculate the critical heat input. Some of them have been discussed below.

Numerical Method for Critical Heat Input Calculation. Numerical method to calculate the critical heat input was presented in Suman *et al.* [29]. They numerically found out the value of heat input so that the value of the radius of curvature at the end of the evaporative section is close to zero. That heat input was termed as the critical heat input.

Analytical Expressions for Critical Heat Input. The analytical expression of the critical heat input in the absence of gravity was derived in Suman and Kumar [32]. For constant heat fluxes in the evaporative and condensing regions $\left(Q'_c = Q/f_1 W_b L \text{ and } Q'_e = Q/[(1-f_2)W_b L]\right)$, the expression for the critical heat input of an MHP in the absence of gravity can be given as

$$Q_{cr} = \frac{2B_1\sigma_1\rho_1\lambda_1 R_o^3}{3B_2L\{1+f_2-f_1\}} \qquad (46)$$

where $B_1 = \left[\{\cot(\alpha+\gamma) - \varphi/2\} + \frac{\cot(\alpha+\gamma)\cos(\alpha+\gamma)\sin\gamma}{\sin\alpha}\right]$

when

$$Q'_c = \frac{Q(m+1)}{W_b(f_1 L)^{m+1}}(f_1 - X^*)^m \quad \text{and} \quad Q'_e = \frac{Q(m+1)}{W_b\left(L(1-f_2)\right)^{m+1}}(X^* - f_2)^m$$

the expression for the critical heat input of an MHP without gravity is obtained as

$$Q_{cr} = \frac{\sigma_1\rho_1\lambda_1 B_1(R_o)^3}{3B_2L(f_2 - f_1) + \dfrac{3B_2L(1-f_2+f_1)}{L^m}\left(\dfrac{m+1}{m+2}\right)} \qquad (47)$$

From the analytical expression we infer the following:

a. Q_{cr} increases with an increase in the latent heat of vaporization.
b. Q_{cr} increases with an increase in the surface tension of a coolant liquid.
c. Q_{cr} increases with an increase in the density of a coolant liquid.
d. Q_{cr} increases with an increase in the width of a groove (also shown by Longtin et al. [18]).
e. Q_{cr} decreases with an increase in the adiabatic section length.
f. Q_{cr} decreases with an increase in the length of a heat pipe.
g. Q_{cr} decreases with an increase in the viscosity of a coolant liquid.

Results and Discussions. The heat pipe inclination had substantial effect on the performance of an MHP. The critical heat input, as calculated, varied significantly with inclination (Table I). It was observed that with an increase in the inclination, the critical heat input decreased due to the opposing (relative to capillary forces, in this configuration) effect of body force (gravity). Therefore, the capillary limit would be reached earlier for an MHP with a higher inclination angle.

The maximum heat transport capacity depended on the length of an MHP. The variation of the critical heat input as a function of length of a V-shaped MHP is given in Table II. It was found that larger an MHP, lesser the critical heat input. This was because that a larger an MHP had more friction loss. Therefore, with an increase in the length of an MHP, the performance of an MHP went down.

TABLE I
Variation of Critical Heat Input (W) with Inclination (°) for a V-Shaped MHP of Length 2 cm [31]

Inclination (°)	Critical heat input (W) $\times 10^2$
0	1.24
10	1.15
30	0.94
45	0.84
60	0.78
90	0.72

TABLE II
Variation of Critical Heat Input (W) with Length (cm) for a V-Shaped MHP [31]

Length (cm)	Critical heat input (W) $\times 10^2$
1	2.4
2	1.15
3	0.71
4	0.51
5	0.385

TABLE III
Variation of the Critical Heat Input with Groove Apex Angle of a V-Shaped MHP of Length 2 cm [31]

Groove apex angle (°)	Critical heat input (W) $\times 10^3$
40	12.8
60	11.5
80	7.5
120	1.2
150	0.08
160	0.026

The apex angle of a V-shaped MHP plays an important role in the capillary pumping. The variation of the critical heat input of a 2-cm-long V-shaped MHP as a function of apex angle (and thus effectively the number of sides) is presented in Table III. The critical heat input decreased with an increase in apex angle. The difference in curvature provided the capillary pumping capacity to an MHP. To pump the same amount of liquid, higher dimensionless curvature difference between the hot and the cold ends was required since the curvature at the cold end ($1/R_o$) decreased with an increase in the apex angle. Moreover, the friction loss increased with an apex angle. The capillary pumping capacity, therefore, decreased with an increase in the apex angle.

b. Dry-Out Length. At the critical heat input, the radius of curvature at the hot end becomes very close to zero. We can say that at the critical heat input, we have zero dry-out length and for any heat input higher than critical heat input, dry-out length increases and we have a dry region in an MHP. This happens because the capillary pumping is not able to sustain the increased rate of evaporation. In the dry region, the coolant liquid is no more available for evaporation and hence two-phase heat transfer does not take place. The length of this region is known as the dry-out length. There are many methods to calculate the dry-out length. Some of them have been discussed below.

Numerical Method for Dry-Out Length. The numerical method to calculate the dry-out length was proposed by Suman and Hoda [29]. The location of the dry-out point can be estimated numerically by calculating the value of X^*, where radius of curvature goes close to zero. The X^* value, where $R^* \to 0$, thus obtained, denotes the nondimensional dry-out length of the heat pipe corresponding to the particular set of process variables.

Analytical Expression for Dry-Out Length. The analytical expression to calculate the dry-out length (L_d) for a set of process variables was presented by Suman and Kumar [32]. The radius of curvature at the hot end for the critical heat input became very close to zero. The capillary pumping became less than the rate of evaporation for any heat input higher than the critical heat. This propagated the dry region starting from the hot end toward the cold end. The dry region of an MHP was known as the dry-out length. If $Q < Q_{cr}$, an MHP was working without generating any dry region. If the heat input was greater than the critical heat input and less than the heat required (Q_2) to onset the dry out was at the junction of the evaporative and adiabatic sections, that is, $R_2^* = 0$, the dry-out region was only in the evaporative section. The general expression for Q_2 was obtained by equating R_2^* to 0 as follows:

$$3B_2'(f_2 - f_1) = 1 - 3B_1' \int_0^{f_1} \left\{ \int_0^{\eta} Q'(\xi) \, d\xi \right\} \quad (48)$$

where $B_2' = B_2 L / \sigma_1 R_o^3 \rho_1 \lambda_1 B_1$.

The dry-out length is calculated by evaluating X_d^* as follows:

$$3B'_3 \int_{f_2}^{X^*_d} \left\{ \int_\eta^1 Q'(\xi) \, d\xi \right\} d\eta = R_2^3 \qquad (49)$$

and

$$L_d = L(1 - X^*_d) \qquad (50)$$

For a constant heat input distribution in the evaporative and condensing sections, the expressions for Q_2 and the dry-out length are given as follows:

$$Q_2 = \frac{\sigma_1 B_1 \rho_1 \lambda_1 R_o^3}{3B_2 L \left\{ (f_2 - f_1) + \frac{(f_1)^2}{2(1-f_2)} \right\}} \qquad (51)$$

$$L_d = L \sqrt{1 - 2 \left\{ \frac{(R_2^*)^3 (1-f_2) \sigma_1 B_1 R_o^3 \rho_1 \lambda_1}{3B_2 LQ} + f_2 - \frac{(f_2)^2}{2} \right\}} \qquad (52)$$

If $Q_2 < Q < Q_1$ (input heat required to onset of dry out at the junction of the adiabatic and condensing sections), the dry region was in the evaporative and adiabatic sections. The expression for Q_1 was obtained by equating R_1^* to zero as follows:

$$1 = 3B'_1 \int_0^{f_1} \left\{ \int_0^\eta Q'(\xi) \, d\xi \right\} d\eta \qquad (53)$$

where $B'_1 = B_2 L^2 W_b / \sigma_1 R_o^3 \rho_1 \lambda_1 B_1$.

The expression for the dry-out length was obtained using Eq. (50), where X^*_d is given as follows:

$$X^*_d = \frac{R_1^3}{3B'_2} + f_1 \qquad (54)$$

For constant heat input distributions in the evaporative and condensing sections, the expressions for Q_1 and the dry-out length are given as follows:

$$Q_1 = \frac{2B_1 \sigma_1 R_o^3 \rho_1 \lambda_1}{3B_2 L f_1} \qquad (55)$$

$$L_d = L \left(1 - \frac{(R_1^*)^3}{3B'_2} + f_1 \right) \qquad (56)$$

If $Q > Q_1$, the dry out was in all the three sections and it is calculated using Eq. (50), where X_d^* is given as follows:

$$3B_1' \int_0^{X_c^*} \left\{ \int_0^\eta Q'(\xi)\, d\xi \right\} d\eta = 1 \qquad (57)$$

For a constant heat input distribution, the expression for dry out is given as follows:

$$L_d = L\left(1 - \sqrt{\frac{2B_1\sigma_1 R_o^3 \rho_1 \lambda_1}{3B_2 L f_1}}\right) \qquad (58)$$

From the analytical expression of dry-out length, we can infer the following:

a. L_d decreases with an increase in the latent heat of vaporization.
b. L_d decreases with an increase in the surface tension of a coolant liquid.
c. L_d decreases with an increase in the density of a coolant liquid.
d. L_d decreases with an increase in the width of a groove.
e. L_d increases with an increase in the adiabatic section length.
f. L_d increases with an increase in the length of a heat pipe.
g. L_d increases with an increase in the viscosity of a coolant liquid.

Results and Discussions. In Fig. 19, it is seen that for every heat input, the value of the dry-out length is more for the rectangular MHP than the triangular MHP. This is due to the fact that a rectangular MHP has higher groove angle than the triangular one. It can also be seen that the dry-out length increases with an increases in heat input. With a higher heat input, higher capillary pressure difference is required and hence the pressure required is much less than the required one. Initially, the dry-out length increases fast, and then it becomes slow with an increase in the heat input. This is because that once the evaporative section is completely dry, the dry-out point has to propagate in the adiabatic section. In this case, the heat has to travel from the evaporative to adiabatic section and therefore, it is slow. We withdraw heat from the condensing section, and hence we cannot have dry out in the condensing section.

2. Sonic Limit

It has been observed that the vapor mass flow rate increases with a decrease in the condenser pressure. Eventually, the vapor velocity reaches the velocity of sound (sonic velocity) at the end of the evaporator. When the

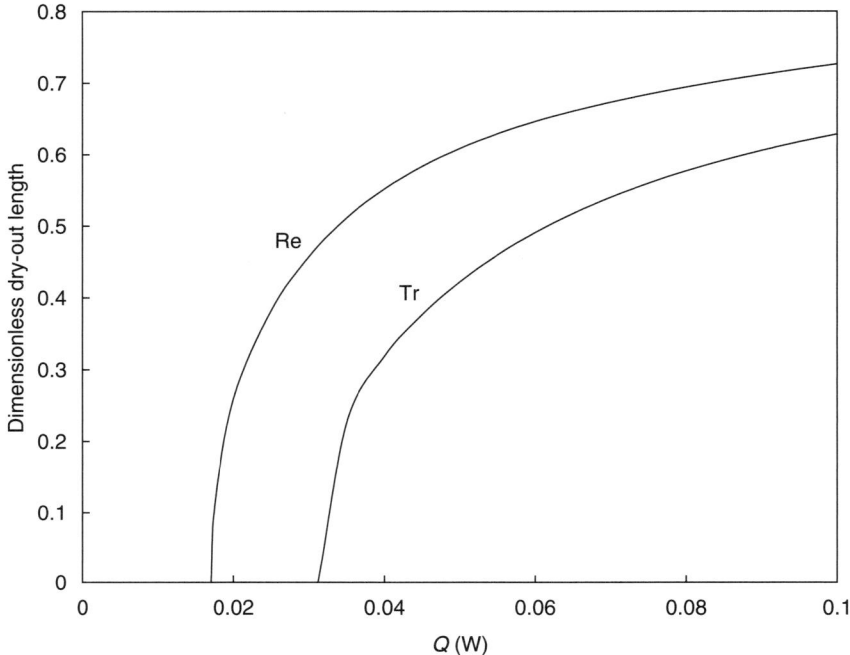

Fig. 19. Variation of the dimensionless dry-out length, with heat input, Q (W), for triangular (Tr), and rectangular (Re) heat pipes.

velocity reaches sonic velocity, further reduction in condenser pressure will not increase the mass flow rate. Hence, the velocity of vapor is restricted to this limit and is called the sonic limit. In the sonic limit region, the vapor flow reaches sonic velocity when exiting heat pipe evaporator, resulting in a constant heat pipe transport power and large temperature gradients. It is because of this power/temperature combination; too much of power is generated at low operating temperature. This is typically only a problem during the start-up of a heat pipe. The heat pipe carries a set power and the large temperature gradient corrects itself as the heat pipe warms up.

3. Flooding Limit

In the operating heat pipes, liquid and vapor flow in opposite directions. The interaction between this countercurrent liquid and vapor flow and the viscous shear forces occurring at the liquid vapor interface may inhibit return of the liquid to evaporator. When this occurs, the heat pipe is said to have reach the flooding limit.

4. Entrainment Limit

When the vapor velocity increases above the flooding limit, it may cause the liquid flow to become unstable. In most severe cases, waves may form and the interfacial shear forces may become greater than the liquid surface tension forces, resulting in an liquid droplet being picked up or entrained by the vapor flow and carried to the condenser. This entrainment of liquid droplets can limit the axial heat flux and this is referred to as the entrainment limit. Both flooding and entrainment result in the excess accumulation of liquid in the condenser region. In the entrainment limit, a high-velocity vapor flow prevents condensate from returning to evaporator since shear forces from high vapor velocities tear liquid and is drawn into vapor flow stream. This is because of heat pipe operating above designed power input or at too low operating temperature. It is observed that the smaller hydraulic diameter elevates this limit.

5. Boiling Limit

When a input heat flux is large, nucleate boiling may occur in the heat pipe. This limitation is known as heat flux or boiling limit. For excessive heater temperatures, film boiling can occur which can limit working fluid circulation. Film boiling in a typical heat pipe evaporator typically initiates at $5-10\,\mathrm{W\,cm^{-2}}$.

6. Viscous Limit

At low operating temperatures, the vapor pressure difference between the evaporative and the condenser region of a heat pipe may be very small. The viscous forces within the vapor region may actually become larger than the pressure gradients caused by the imposed temperature field. When this occurs, the pressure gradients within the vapor region may not be sufficient to generate flow and vapor may become stagnant. The no-flow and low-flow condition in the vapor portion of a heat pipe is referred to as its viscous limit. It is most often observed in cryogenic heat pipe. Viscous forces prevent vapor flow in the heat pipe since viscous forces on vapor reduce vapor pressure at condenser to zero. This is because that the heat pipe operates below recommended operating temperature. At very low temperatures, the vapor pressure difference between the condenser and evaporator can be so small that viscous forces in the vapor core are dominant and limit the vapor flow. Its potential solution is to increase heat pipe operating temperature or find alternative working fluid.

7. Condenser Limit

Cooling ability of the condenser is possibly affected by the presence of noncondensable gases. The presence of even a small quantity of a noncondensable gas in the condensing vapor has a profound influence on the resistance to heat transfer in the region of the liquid–vapor interface. This noncondensable gas is carried with the vapor toward the interface where it gets accumulated. The partial pressure of the gas at the interface becomes higher than that at the bulk, producing a driving force for gas diffusion away from the surface. This motion is exactly counterbalanced by the motion of the vapor gas toward the surface. Since their sum remains constant, the partial pressure of the vapor at the interface is less than that at the bulk, providing the driving force for vapor diffusion toward the interface.

8. Vapor Continuum Limit

If the vapor flow is such that it reaches its continuum limit because it is rarefied, the molecules collide with a wall more frequently that among themselves, and therefore the property of the vapor changes. This limit needs to be investigated.

9. Frozen Start-up Limit

This is a start-up problem when a heat pipe has to start from a frozen condition. It happens that the vapor from evaporation is refrozen and depletes fluid available in the evaporation zone. It is well documented in Ref. [12].

VII. Future Direction

MHPs have become one of the most promising cooling devices because of their high efficiency, reliability, and cost-effectiveness. Although the use of MHPs for enhanced heat transfer is becoming more common, the exact nature of liquid evaporation from the corners of an MHP and the associated capillary pumping capacity has not been investigated in detail. Some important aspects to be investigated in future have been discussed below.

In an MHP, a vapor flow is important and sometime will dominate the performance of the MHP. Vapor shear can enable effective continuous condensation in the microgravity environment. What is the effect of vapor shear on the friction on the interface between vapor and liquid film?

Proper analysis of liquid–vapor interfaces in an operational MHP helps in predicting the real heat pipe limitations. The work of Beavers and Josef [154] was one of the first attempts to study the fluid flow boundary conditions at the interface region. They performed experiments and detected a slip in the velocity at the interface. The gradient in the liquid pressure, vapor pressure, and temperature differences at the liquid–vapor interface, control fluid flow, and phase change heat transfer processes in pure microscale liquid–vapor systems, like an evaporating extended meniscus (e.g., theoretically discussed in Refs. [155–167] and experimentally discussed in Refs. [168–179]). Wayner and his coworkers published two seminal papers: a basic one [155] and an applied one [156]. In his work [156], they have demonstrated and predicted the configuration used in many heat pipes today. It also demonstrated that an understanding of the basic interfacial physics of the evaporative meniscus is needed to design such systems. Jing and Carey [180] explored a new efficient computational method for predicting heat and mass transfer near liquid–vapor interface in a microbubble heat pipe.

However, the basic thermophysical principles controlling these small capillary pressure systems are not well understood. Basu et al. [181] initiated a basic research program to rectify this deficiency on a generic system, which they call the constrained vapor bubble (CVB). The current CVB can be viewed as a large-scale version of an MNP with a large Bond number in the earth environment but a small Bond number in microgravity. The most important limitation that is directly related to the liquid–vapor interface is the entrainment limitation. The phenomenon of entrainment occurs when high-velocity vapor flow passes over the liquid–vapor interface and entrains the small liquid drops from the interface. Investigation of the flow pattern for the two-phase flow in microchannels will be helpful for the understanding an MHP. There has been a lot of experimental works on two phase flow pattern in microchannels [182–186].

Many researchers have studied the entrainment limit in conventional heat pipes [187], but the phenomenon of entrainment seems to be more complex and the previous theoretical and experimental results are limited to a number of configurations, wick structures, and materials. The results of the previous works include a number of correlations, criteria, and simulations. The most accurate experimental result corresponds to an attractive approach followed by Kim and Peterson [188]. They reviewed 12 important models and completed extensive experiments with a copper–water, rectangular cross section, and 2.16-m-long heat pipe. They also performed parametric studies and used computer modeling to compare their results. They classified the entrainment in their heat pipe experiment into three major modes: wave-induced, pulsating, and intermediate modes according to the flow visualization experiment and the temperature fluctuation patterns at the onset of the individual entrainment modes. Finally, they proposed a correlation for predicting the critical Weber number for the stability of the liquid interface. The other important result of the experiments done by Kim and Peterson [188] was that the capillary limit occurred before the boiling or entrainment limit. It can be found from the previous research that for the analysis of liquid–vapor interface in an MHP, both experimental and numerical analyses are required.

Many steady-state and transient models have been developed. But, they have been developed based on a number of assumptions. Some of those assumptions are not valid in real situations. For example, most of the models are 1D in nature, and they take variations of radius of curvature perpendicular to the flow. But, the variation of the radius of curvature along the length might be important to be considered, as the length is very small. The 2D flow and radius of curvature analysis will not only help us to understand capillary phenomenon in detail, but will also modify the sensitivity analysis results for the variation of contact angle and surface tension. These models have also considered constant vapor pressure and negligible shear liquid–vapor interface shear. Due to small flow area, vapor pressure will vary and shear at the liquid vapor interface will not be negligible. The correct value of friction factor is also important since flow of fluid in an MHP strongly depends on the friction factor. Moreover, the friction factor at microlevel is different from macrolevel. Some researchers [100,135–141] have presented an expression for friction factor. But the general agreement is yet to be achieved. There are a few experimental studies for model validation, and therefore we should have experimental results to validate these mathematical models and to understand the capillary phenomenon. In most of mathematical models [26–33], the evaporative heat fluxes have been assumed which is not true as the evaporative heat flux will be defined by the liquid temperature and the vapor pressure. In some studies [8], the evaporative heat flux as a function of vapor pressure and liquid temperature have been presented, but they are for only for the evaporative section and hence they should be extended to all three sections. The heat transfer in the evaporative section has been well investigated [189–194]. But, in an MHP we also have a condensing section where condensation takes place. Therefore, detailed study of condensation in the condensing section should be done. There has been some model research on boiling and condensation of a two-phase flow in a triangular microchannel [195–197].

Most of the works reported in the literature focus on the flow in a single channel. But, the heat transfer capacity of a single channel is small, and thus for a practical application, we need to have multiple grooves on a substrate. Each groove will act as a separate MHP. One idea is to seal the substrate from the top and let each groove operate separately. The present state of model is applicable for this kind of system. In this case, we want to increase the vapor flow area by allowing the grooves to interact among themselves.

For such a situation, the interaction among the channels may be important. The situation may arise when flooding in some channels and dry out in others.

Suman [134] has shown that the surface tension gradient has a significant effect on the liquid flow in an MHP. In the study, the surface tension gradient was generated due to the temperature difference between the hot and the cold ends which opposed the flow. But it can be changed such that it can facilitate the liquid flow from the cold end to the hot ends using surfactants [198], which will improve the performance of an MHP.

In an MHP, the flow of liquid is due to the difference in the radius of curvature between the hot and the cold ends. Its heat transport capacity is less than that of a wicked heat pipe. But its simple design and direct integration on the substrate make it suitable for many applications. Heat transport capacity can be increased by cascading the grooves. It has been attempted and observed that the heat transport capacity can be significantly improved by using an electric field. By applying an electric field at the liquid–vapor interface, pressure difference can be increased if the coolant liquid is dielectric in nature. This research is based on the assumption that both augmentation of the heat transport capacity and active thermal control of MHPs can be achieved through the application of a static electric field. When an electric field is strategically applied within MHPs at the liquid–vapor interface that are internally present, dielectrophoretic EHD forces can be induced due to the discontinuity of the electrical permittivity of the medium. These forces can contribute to a pressure jump condition at the interface which can effectively reduce the liquid pressure in the region affected by the field. This pressure reduction can be used to augment the flow of liquid from the condenser to the evaporator region [110]. Employing an electric field, we can increase or decrease the substrate temperature, and hence the device temperature by varying the field strength. But, these models [125] are semi empirical in nature. The effect of an electrical field has not been directly incorporated into flow of fluid. Therefore, a model developed from the first principle and its experimental validation is required to understand the effect of an electrical field in the MHPs. Such an attempt has been made by Suman [126].

Theory and fundamental phenomenon that cause each of heat pipe limitations has been the subject of a considerable number of investigations for conventional heat pipes. The few famous works have been presented and discussed by Dunn and Reay [9], Brennan and Kroliczek [199], Chi [8], Cotter [200], Chisholm [201], Tien [202], and Feldman [203]. However, out of the numerous works during the past 40 years, only a limited number of works concern the operation limitations in an MHP [204]. These limitations have to be studied for an MHP and they might need to be reformulated.

The entrainment limitation is more likely to change since in an MHP as we do not have wick structure rather there is flow of fluid in a groove.

The microgroove can be coated in the evaporative section with a catalyst, which promotes an endothermic reaction, and another catalyst can be coated in the condenser section, which promotes the reverse reaction of that, which occurs in the evaporative section. The heat capacity of the MHP can be dramatically improved by using this technique. Presently, we transfer the latent heat of vaporization, and in that case, the amount of heat transfer will depend on the heat of reactions. A similar study is done by Riabov and Provotorov [206]. When the heat of reaction is higher than the latent heat, the performance will be enhanced. But, the issue is to select such a catalyst and reaction, and then overcome the fabrication and coating challenges.

An important advancement in the field of biological applications is to develop MHP heat spreader which can be implanted to arrest epileptic seizures by rapid cooling of small regions of the brain. Research is underway in developing such MHPs with metal foils and polymers in the Peterson and his coworkers groups. An increase in the electrical activity of the brain causes seizure and differential cooling of a portion of the brain can reduce the electrical activity. The challenge in this is that a small region of the brain should be cooled without significantly increasing the temperature of any other portion. Another biomedical application of MHPs is a heat pipe catheter of the size of a hypodermic needle, which assures constant temperature operation within a therapeutic temperature range for use in hyperthermia cancer treatments and may be used to treat cancerous tumors in body regions previously untreatable. Other biological applications are possible and there is a need to find such problems and solve them using MHPs.

Nanotechnology is an emerging field. A serious effort has been made to understand its thermal science and technology. Thermal problem is bound to arise in nanotechnology [206,207]. There are typically two types of problems. One is the management of heat generated in nanoscale devices to maintain the functionality and the reliability of those devices. The other is to utilize nanostructures to manipulate the heat flow and energy conversion. Examples of the thermal management of nanodevices are the heating issues in integrated circuits [208] and in semiconductor lasers [209]. Examples in the manipulation of heat flow and energy conversion include nanostructures for thermoelectric energy conversion [210,211], thermophotovoltaic power generation [212], and data storage [213]. Heat transfer at nanoscale may differ significantly from that at macro- and microscales. With device or structure characteristic length scales becoming comparable to the mean free path and wavelength of heat carriers (electrons, photons, phonons, and molecules), classical laws are no longer valid and new approaches must be taken to predict heat transfer at nanoscale [214,215]. Well-known examples are the failure of Fourier law to predict the thermal conductivity of composite nanostructures such as superlattices [215–217] and the failure of the Stefan–Boltzmann law in predicting radiation heat transfer across small gaps [218,219]. Although much work has been done in this area, there is still an immediate need for a better understanding of thermal phenomena in nanostructures.

Heat pipes at the nanolevel can be termed as the "nano heat pipe," and an extension of an MHP to nanolevel is expected to be a very promising area for research. A nano heat pipe will be different from an MHP's modeling and experiments. It is the uniqueness of nanostructures regarding their nanoscale thermal properties that is emphasized here, by giving examples of thermal phenomena not present in typical bulk materials [220]. Dresselhaus' comment on two unique and very different aspects of heat transport at interfaces in nanostructures also depend on length scales, involve interesting fundamental science issues, and perhaps may give rise to useful applications some day. For such a structure, the thermal conductivity in the direction of the nanotube axis can be very high, even higher than that of the best 3D thermal conductor [205]. This exceptionally high thermal conductivity arises largely because of the unique properties of nanostructures: (1) the carbon–carbon bond is the strongest chemical bond in nature, (2) the interfaces with the vacuum, along with a restricted number of final states to which scattering can occur, severely limits phonon–phonon scattering, (3) the sizable semiconductor band gap virtually eliminates electron–phonon scattering, and (4) the high structural perfection that can be achieved severely limits the defect scattering mechanism. While graphite in-plane (along with diamond) has the highest thermal conductivity of known materials, the unique properties of the nanotube, enumerated above, further eliminates many of the possible scattering channels that occur in graphite, thereby allowing a single isolated carbon nanotube to behave like a wave guide for heat transfer [221]. Can such heat pipes be utilized in nanosystems, and can some of the unique properties of nano heat pipes be used to design other structures at the nanoscale with enhanced thermal transport?

VIII. Conclusions

The MHP is an effective device in various heat transfer, heat recovery, and heat exchanger applications, including biological and nanotechnology applications because of their high efficiency, reliability, capability of working at small length scale, and cost-effectiveness. The use of MHPs is particularly promising, as an increase in power density requires high heat flux removal possibly to be met by the phase change heat transfer for temperature control.

Although the use of MHPs for enhanced heat transfer is becoming more common, the exact nature of liquid evaporation from the corners of an MHP and the associated capillary pumping capacity has not been fully understood. The typical steady-state and transient models for an MHP are able to capture some of its behavior. However, these models use several assumptions. These assumptions need to be removed and a detailed heat transfer model coupled with fluid flow for heat pipes is an immediate need with a proper consideration of thermodynamics at the liquid–vapor interface along with fluid and mass transfer in the microchannel. For a design of the grooved heat pipe, along with mathematical models, we should understand various limitations. These limitations on the heat transport capacity of an MHP are similar to those of a conventional heat pipe. During a steady-state operation, along with the capillary limit, several other important mechanisms can limit the maximum amount of heat that an MHP can transfer. Among these are the viscous, sonic, entrainment, and boiling limits. An exact mathematical modeling and prediction of these limitations need to be studied for a particular MHP as the liquid meniscus in MHP is well defined. Due to further miniaturization, we started thinking about nanoscale devices; thus, the heat transfer problem is bond to occur. As a result, we need to prepare ourselves for such problems associated with nanotechnology, biotechnology, and current energy problems. A natural choice could be to extend an MHP concept (or similar concept) to the nanoscale. Thus, the motivation of this review is to summarize the state of the art in the MHP and leave us with open challenges associated with heat transfer problems in nanotechnology, biotechnology, and universal energy crisis.

Acknowledgments

The invaluable suggestions throughout the manuscript preparation from Prof. G. P. "Bud" Peterson (also for providing the original figures of his papers) of the Georgia Institute of Technology, Atlanta, Georgia, and the correspondence with Prof. P. C. Wayner Jr. of the Rensselaer Polytechnic Institute on the evaporative heat flux research are gratefully acknowledged. The author would like to acknowledge Profs. Sunando DasGupta and Sirshendu De of IIT Kharagpur, India, for introducing this topic, and under their guidance, he conceptualized many of the ideas which formed the basis of this chapter. The author thanks his coauthors who enriched the author's understanding on this topic, and to Kiran Mishra and Prabhat Kumar for their help in manuscript preparation. Lastly, this would not be possible without the voluminous research that has been produced by the authors of the referred articles in this chapter.

Nomenclature

GREEK SYMBOLS

α	half apex angle of polygon (radian)
β	inclination of substrate with horizontal (radian and geometric factor)
γ	contact angle (radian)
ϕ	curvature (m^{-1})
λ_l	latent heat of vaporization of coolant liquid (J kg^{-1})
μ_l	viscosity of coolant liquid (kg m^{-1} s^{-1})
ρ_l	density of coolant liquid (kg m^{-3})
ρ_g	density of vapor (kg m^{-3})
σ_l	surface tension of coolant liquid (N m^{-1})
τ_w	wall shear stress (N m^{-1})
η, ξ, ζ	variables used for integration

SYMBOLS

a	side of V groove
A_{cs}	area of cross section for fluid flow (m^2)
A_l	total liquid area (m^2)
A_g	total vapor area (m^2)
A_t	total cross-sectional area of heat pipe
B	performance factor
$B_1, B'_1, B'_2, B'_3, B_2$	constants
C_{pl}	specific heat of coolant liquid (J kg^{-1} °C^{-1})
C_{ps}	specific heat of substrate (J kg^{-1} °C^{-1})
f	friction factor
f_1	nondimensional coordinate of junction of condensing and adiabatic sections
f_2	nondimensional coordinate of junction of adiabatic and evaporative sections
g	acceleration due to gravity (m s^{-2})
$H(l)$	integral of axial heat transport fraction over L
K'	constant in expression for B_2
K_1^+ and K_1^-	flow shape factors
K_s	thermal conductivity of substrate (W m^{-1} K^{-1})
K_l	thermal conductivity of coolant liquid (W m^{-1} K^{-1})
L	length of heat pipe (m)
L_d	dry-out length (m)
L_h	half of total wetted length (m)
m	constant in expression for generalized heat flux
N_{Re}	Reynolds number
P_l	liquid pressure (N m^{-2})
P_l^*	nondimensional liquid pressure
P_R	reference pressure (N m^{-2})
P_{vo}	pressure in vapor region (N m^{-2})
Q	net heat flux supplied (W m^{-2})
Q_{ss}	heat flux supplied at steady state (W m^{-2})
Q'	heat input (W)
Q_{in}	reference heat input (W)
Q'_c	heat flux in condensing region (W m^{-2})
Q_{cr}	critical heat flux (W m^{-2})
Q'_e	heat flux in evaporative region (W m^{-2})
Q_v	heat flux for vaporization of liquid, W m^{-2}
Q_1	heat flux at solid–liquid interface (W m^{-2})
R	radius of curvature (m)
R^*	nondimensional radius of curvature
R_1^*	nondimensional radius of curvature at f_1
R_2^*	nondimensional radius of curvature at f_2
R_o	reference radius (m)
Re	rectangle
R_l	meniscus surface area per unit length (m)
R_{th}	thermal resistance (W K^{-1} m^{-2})
T_{con}	temperature at cold end (°C)
Tr	triangle
T_R	reference temperature (°C)
T_s	temperature of substrate (°C)

T_s^*	dimensionless substrate temperature	X^*	nondimensional coordinate along heat pipe
v_l	specific volume of liquid	X_d^*	nondimensional coordinate of onset of dry region
v_g	specific volume of gas		
V_g	axial vapor velocity (m s^{-1})		
V_l	axial liquid velocity (m s^{-1})	\multicolumn{2}{l	}{SUBSCRIPTS}
V_g^*	nondimensional vapor velocity		
V_l^*	nondimensional liquid velocity	c	condenser section
V_R	reference liquid velocity (m s^{-1})	d	onset of dry-out region
W_b	perimeter of heat pipe polygon (m)	e	evaporative section
		g	vapor
x	coordinate along heat pipe (m)	l	coolant liquid
		s	substrate

References

1. King, C. R. (1931). *Engineer* **152**, 405–406.
2. Perkins, L. P. and Buck, W. E. (1892). Improvements in Devices for the Diffusion or Transference of Heat, United Kingdom Patent No. 22,272.
3. Gay, F. W. (1929). Heat Transfer Means, United States Patent No. 1,725,906.
4. Gaugler, R. (1944). Heat Transfer Device, United States Patent No. 2,350,348.
5. Trefethen, L. (1962). On the Surface Tension Pumping of Liquids or a Possible Role of the Candlewick in Space Exploration, G. E. Tech Info., Serial No. 615 D114.
6. Grover, G. (1966). Evaporation – Condensation Heat Transfer Device, United Sates Patent No. 3,229,759, Application filed on December 2, 1963, Approved January 18.
7. Grover, G. M., Cotter, T. P., and Erikson, G. F. (1964). *J. Appl. Phys.* **35**, 1190–1191.
8. Chi, S. W. (1976). "Heat Pipe Theory and Practice." Hemisphere, Washington, DC, USA.
9. Dunn, P. D. and Reay, D. A. (1982). "Heat Pipes." Pergamon, Oxford, UK.
10. Terpstra, M. and Van Veen, J. G. (1987). "Heat Pipes: Construction and Application." Elsevier Applied Science, New York, NY, USA.
11. Peterson, G. P. (1994). "An Introduction to Heat Pipes: Modeling, Testing, and Applications." John Wiley, New York, NY, USA.
12. Faghri, A. (1995). "Heat Pipe Science and Technology." Taylor & Francis, Washington, DC, USA.
13. Peterson, G. P. (2001). "The MEMS Handbook." Chapter 31, CRC Press, Boca Raton, FL, USA.
14. Mallik, A. K. and Peterson, G. P. (1995). *J. Electron. Packag.* **117**, 82–87.
15. Mallik, A. K., Peterson, G. P., and Weichold, M. H. (1992). *J. Electron. Packag.* **114**, 436–442.
16. Peterson, G. P., Duncan, A. B., and Weichold, M. H. (1993). *J. Heat Transf.* **115**, 750–756.
17. Babin, B. R., Peterson, G. P., and Wu, D. (1990). *J. Heat Transf.* **112**, 595–601.
18. Longtin, J. P., Badran, B., and Gerner, F. M. (1994). *J. Heat Transf.* **116**, 709–715.
19. Wu, D. and Peterson, G. P. (1991). *J. Thermophys. Heat Transf.* **5**(2), 129–134.
20. Peterson, G. P. and Ma, H. B. (1996). *J. Heat Transf.* **118**, 731–739.
21. Khrustalev, D., Faghri, A. (1994). *J. Heat Transf.* **116**(1), 189–198.
22. Xu, X. and Carey, V. P. (1990). *J. Thermophys. Heat Transf.* **4**, 512–520.
23. Ravikumar, M. and DasGupta, S. (1997). *Chem. Eng. Commun.* **160**, 225–248.
24. Ma, H. B. and Peterson, G. P. (1996). *J. Heat Transf.* **118**, 740–746.
25. Khan, A.Md., Mishro, S., De, S., and DasGupta, S. (1999). *Int. J. Trans. Phenom.* **1**, 277–289.

26. Ha, J. M. and Peterson, G. P. (1998). *J. Heat Transf.* **120**, 452–457.
27. Anand, S., De, S., and DasGupta, S. (2002). *Int. J. Heat Mass Transf.* **45**, 1535–1543.
28. Catton, I. and Stroes, G. R. (2002). *J. Heat Transf.* **124**, 162–168.
29. Suman, B., De, S., and DasGupta, S. (2005). *Int. J. Heat Fluid Flow* **26**(3), 495–505.
30. Suman, B., De, S., and DasGupta, S. (2005). *Int. J. Heat Mass Transf.* **48**(8), 1633–1646.
31. Suman, B. and Hoda, N. (2005). *Int. J. Heat Mass Transf.* **48**, 2090–2101.
32. Suman, B. and Kumar, P. (2005). *Int. J. Heat Mass Transf.* **48**, 4498–4509.
33. Suman, B. (2006). On the fill charge and sensitivity analysis of a V-shaped micro heat pipe. *AIChE J.* **52**, 3041–3054.
34. Suman, B. and Hoda, N. (2007) On the transient analysis of a V-shaped microgrooved heat pipe. *J. Heat Transf.* **129**(11), 1584–1591.
35. Peterson, G. P. (1993). Investigation of micro heat pipes fabricated as an integral part of silicon wafers. In "Advances in Heat Pipe Science and Technology, Proceedings of the International Heat Pipe Conference – 8th, Beijing, September 14–18, 1992" (T. Ma, ed.), pp. 385–395. International Academic Publishers, Beijing, People's Republic of China, CODEN: 61QQA7 Conference.
36. Itoh, A., Polasek, F., Itoh, R., and Lab, D. (1993). Development and application of micro heat pipes. In "Heat Pipe Technology, Proceedings of the 7th International Heat Pipe Conference, Minsk, May 21–25, 1990" (L. L. Vasiliev, ed.), pp. 1295–1310. Begell House, New York, NY, USA, CODEN: 65TSAM Conference.
37. Peterson, G. P. (1992). *Appl. Mech. Rev.* **45**, 175–189.
38. Peterson, G. P. (1996). *Appl. Mech. Rev.* **49**(10, Pt 2), 175–183.
39. Suman, B. (2007). Modeling, experiment, and fabrication of micro-grooved heat pipes. *Appl. Mech. Rev.* **60**(3), 107–120.
40. Cao, Y., Faghri, A. (1994). *J. Enhanced Heat Transf.* **1**(3), 265–274.
41. Shyu, R. J. (1997). Heat pipe research and applications in Taiwan. In "Heat Pipe Technology: Theory, Applications and Prospects, Proceedings of the 5th International Heat Pipe Symposium, Melbourne, November 17–20, 1996" (J. Andrews, A. Akbarzadeh, and I. Sauciuc, eds.), pp. 55–65. Elsevier, Oxford, UK, CODEN: 66GKA8 Conference.
42. Peterson, G. P., Swanson, L. W., and Gerner, F. M. (1998). Micro heat pipes. In "Microscale Energy Transport" (C.-L. Tien, A. Majumdar, and F. M. Gerner, eds.), pp. 295–337. Taylor & Francis, Washington, DC, USA, CODEN: 66DPA8 Conference; General Review written in English. CAN 129:5903 AN 1998:348498 CAPLUS.
43. Spalding, D. B. (1980). Mathematical modeling of fluid-mechanics heat transfer and chemical-reaction processes, a lecture course, CFDU Report HTS/80/1, Imperial College, London, UK.
44. Ivanovskii, M. N., Sorokin, V. P., and Yagodkin, I. V. (1982). "The Physical Principles of Heat Pipes." Oxford University Press, Oxford, UK.
45. Yuan, S. W. and Finkelstein, A. B. (1955). Laminar flow with injection and suction through a porous wall. In "Proceedings of Heat Transfer and Fluid Mechanics Institute." University of California, Los Angeles, CA, USA.
46. Mathews, J. H. (1992). "Numerical Methods for Mathematics, Science and Engineering." 2nd edn. Prentice-Hall, Inc, Upper Saddle River, NJ, USA.
47. Nouri-Borujerdi, A. and Layeghi, M. (2005). *Heat Transf. Eng.* **26**, 45–58.
48. Cotter, T. P. (1984). Principles and prospects of micro heat pipes. In "Proceedings of the 5th International Heat Pipe Conference." pp. 328–332. Tsukuba, Japan.
49. Cao, Y., Faghri, A., and Mahefkey, E. T. (1993). *ASME HTD* **236**, 55–62.
50. Tuckerman, D. B. and Pease, R. F.W. (1981). *IEEE Electron Device Lett.* **2**(5), 126–129.
51. Take, K. and Webb, R. (2001). *J. Electron. Package.* **123**, 189–195.
52. Ortega, A., Lall, B. S., Chicci, J., Aghazadeh, M., and Kiang, B. (1993). Heat transfer in low aspect ratio horizontal enclosure for laptop computer application. In "Semiconductor

Thermal Measurement and Management Symposium – SEMI-THERM, February 2–4, 1993." pp. 42–49. IEEE, Austin, TX.
53. Xie, H., Aghazadeh, M., Lui, W., and Haley, K. (1995). Thermal solutions to pentium processors in TCP in notebooks and sub-notebook. *In* "Proceedings of the 1995 Electronic Components and Technology Conference (ECTC), May 21–24, 1995." pp. 201–210. IEEE, Las Vegas, NV.
54. Iswannath, R. and Ali, I. A. (1995). Thermal modeling of high performance packages in portable computers. *In* "Proceedings of the 1995 Electronic Components and Technology Conference (ECTC), May 21–24, 1995." pp. 1122–1133. IEEE, Las Vegas, NV.
55. Xie, H., Aghazadeh, M., and Toth, J. (1995). The use of heat pipes in the cooling of portables with high power packages – A case study with the Pentium processor-based notebooks and sub-notebooks. *In* "Proceedings of the 1995 Electronic Components and Technology Conference (ECTC), May 21–24, 1995." pp. 906–913. IEEE, Las Vegas, NV.
56. Marongiu, M. J., Kusha, B., Fallon, G. S., and Watwe, A. A. (1998). Enhancement of multi chip modules (MCMs) cooling incorporating micro-heat pipes and other high thermal conductivity materials into mechanical micro channel heat sinks. *In* "Proceeding of the 1995 Electronic Components and Technology Conference (ECTC), May 21–24, 1998." pp. 1122–1133. IEEE, Las Vegas, NV.
57. Adkins, D. R., Shen, D. S., Palmer, D. W., and Tuck, M. R. (1994). Silicon heat pipes for cooling electronics. *In* "Proceedings of the 1st Annual Spacecraft Thermal Control Symposium, November 16–18, 1994." p. 10. Albuquerque, NM, USA.
58. Weichold, M. H., Peterson, G. P., and Mallik, A. K. (1993). Vapor Deposited Micro Heat Pipes, United States Patent No. 5,179,043. January 12.
59. Kojima, Y., Yamazaki, N., Yoshida, K., Mishiro, H., and Murakami, M. (1993). LSI cooling system with micro heat pipe. *In* "Advances in Heat Pipe Science and Technology, Proceedings of the International Heat Pipe Conference – 8th, Beijing, September 14–18, 1992" (T. Ma, ed.), pp. 539–542. International Academic Publishers, Beijing, People's Republic of China, CODEN: 61QQA7 Conference written in English. CAN 123:303291 AN 1995:829767 CAPLUS.
60. Peterson, G. P. and Oktay, S. (1990). A Bellows Heat Pipe, United States Patent No. 4,951,740 issued August 28.
61. Peterson, G. P. and Oktay, S. (1997). Coupled Flux Transformer Heat Pipes, U.S. Patent No. 5,647,429, January 15.
62. North, M. T. and Avedisian, C. T. (1993,) *J. Electron. Packag.* **115**(1), 112.
63. Rosenfeld, J. H., Gernet, N. J., and North, M. T. (1994). *ASME HTD* **273**, 93–100.
64. Fletcher, L. S. and Peterson, G. P. (1993). Micro Heat Pipe Catheter for Local Tumor Hyperthermia, United States Patent No. 5,190,539, March 2.
65. Peterson, G. P. and Fletcher, L. S. (1995). Temperature Control Mechanisms for Micro Heat Pipe Catheters, United States Patent No. 5,417,686, May 23.
66. Peterson, G. P. and Fletcher, L. S. (1997). Treatment Method Using a Micro Heat Pipe Catheter, United States Patent No. 5,591,162, January 7.
67. Camarda, C. J., Peterson, G. P., and Rummler, D. R. (1996). Micro Heat Pipe Panels and Method for Producing Same, United States Patent No.5,527,588, June 18.
68. Camarda, C. J., Peterson, G. P., and Rummler, D. R. (1996). Micro Heat Pipe Panels and Method for Producing Same, U.S. Patent No. 5,527,588, June 18.
69. Peterson, G. P. (2001). "US Government-Authored or -Collected Reports." Written by G. P. Peterson. Report Date: May 30, Report No. A223004.
70. Baker, K. W., Dustin, M. O., and Crane, R. (1990). Heat receiver design for solar dynamic space power systems. *In* "International Solar Energy Conference, April 1–4, 1990." Miami, FL, USA.

71. Andraka, C. E., Diver, R. B., and Wolf, D. A. (1992). Design, fabrication, and testing of a 30 kW screen-wick heat-pipe solar receiver. *In* "27th Intersociety Energy Conversion Engineering Conference, August 3–7, 1992." San Diego, CA, USA.
72. Adkins, D. R. (1993). High-flux testing of heat pipes for point-focus solar collector systems. *In* "National Conference and Exposition on Heat Transfer, August 8–11, 1993." Atlanta, GA, USA.
73. Boman, B. L. and Elias, T. I. (1990). Tests on a sodium/Hastelloy X wing leading edge heat pipe for hypersonic vehicles. *In* "5th AIAA and ASME Joint Thermophysics and Heat Transfer Conference, June 18–20, 1990." Seattle, WA, USA.
74. Cao, Y. and Faghri, A. (1992). *Heat Pipes Thermosyphons ASME HTD* **221**, 43–52.
75. Cao, Y. and Faghri, A. (1993). *J. Heat Transf.* **115**(3), 819.
76. Gottschlich, J. M. and Meininger, M. (1992). Heat pipe turbine vane cooling. *In* "27th Intersociety Energy Conversion Engineering Conference, August 3–7, 1992." San Diego, CA, USA.
77. Faghri, A. and Guo Z. (2005). *Int. J. Heat Mass Transf.* **48**(19–20), 3891–3920.
78. Faghri, A. (2005). Micro Heat Pipe Embedded Bipolar Plate for Fuel Cell Stacks, U.S. Patent No. 2005/0026015 A1.
79. Chi, S. W. (1976). "Heat Pipe Theory and Practice." McGraw-Hill, New York, NY, USA.
80. Gerner, F. M., Longtin, J. P., Henderson, H. T., Hsieh, W. M., Ramadas, P., and Chang, W. S. (1992). *ASME HTD* **206**(3), 99–104.
81. Longtin, J. P., Badran, B., and Gerner, F. M. (1993). A one-dimensional model of a micro heat pipe during steady-state operation. *In* "Advances in Heat Pipe Science and Technology, Proceedings of the International Heat Pipe Conference – 8th, Beijing, September 14–18, 1992" (T. Ma, ed.), pp. 401–407. International Academic Publishers, Beijing, People's Republic of China, CODEN: 61QQA7.
82. Longtin, J. P., Badran, B., and Gerner, F. M. (1992). *ASME HTD* **200**, 23–33, CODEN: ASMHD8 ISSN: 0272–5673.
83. Khrustalev, D. and Faghri, A. (1995). *J. Heat Transf.* **17**, 1048–1054.
84. Wang, C. Y., Groll, M., Rosler, S., and Tu, C. J. (1994). *Heat Recovery Syst. CHP* **14**(4), 377–389.
85. Peterson, G. P. and Ma, H. B. (1999). *J. Heat Transf. Trans. ASME* **121**(2), 438–445.
86. Peterson, G. P. and Ma, H. B. (1998). *ASME HTD* **357**(3), 233–242, CODEN: ASMHD8 ISSN: 0272–5673.
87. Sartre, V., Zaghdoudi, M. C. and Lallemand, M. (2000). *Int. J. Thermal Sci.* **39**(4), 498–504.
88. Tio, K. K., Liu, C. Y., and Toh, K. C. (2000). *Heat Mass Transf.* **36**(1) 21–28.
89. Riffat, S. B., Zhao, X., and Doherty, P. S. (2002). *Appl. Thermal Eng.* **22**, 1047–1068.
90. Collier, J. G. (1981). "Convective Boiling and Condensation." Mc-Graw-Hill, New York, NY, USA.
91. Colwell, G. T. and Chang, W. S (1984). *Int. J. Heat Mass Transf.* **27**(4) 541–551.
92. Zhang, J. and Wong, H. (2002). *ASME HTD* **372**(4), 221–224, CODEN: ASMHD8 ISSN: 0272–5673. *J. Written Eng.* CAN 139:8608 AN 2003:442237 CAPLUS.
93. Kalahasti, S. and Joshi, Y. K. (2002). *Trans. Compon. Package. Technol.* **25**(2), 554–560.
94. Kim, S. J., Seo, J. K., and Do, K. H. (2002). *Int. J. Heat Mass Transf.* **45**, 1535–1543.
95. Khandekar, S., Groll, M., Luckchoura, V., Findl, W., and Zhuang, J. (2003). Micro heat pipes for stacked 3D microelectronics modules. *In* "Advances in Electronic Packaging 2003, Proceedings of the International Electronic Packaging Technical Conference and Exhibition, July 6–11, 2003." Vol. 2, pp. 245–251. Maui, HI, USA, CODEN: 69GNMG CAN 142:473969 AN 2005:175853 CAPLUS.

96. Do, K. H., Kim, S. J., and Hwang, G. (2003). Modeling and thermal optimization of a micro heat pipe with curved triangular grooves. *In* "Advances in Electronic Packaging 2003, Proceedings of the International Electronic Packaging Technical Conference and Exhibition, July 6–11, 2003." Vol. 2, pp. 271–278. Maui, HI, USA.
97. Suh, J. S. and Park, Y. S. (2003). *Tamkang J. Sci. Eng.* **6**(4), 201–206.
98. Schneider, G. E. and DeVos, R. (1980). Nondimensional Analysis for the Heat Transport Capability of Axially-Grooved Heat Pipes Including Liquid/vapor Interaction, AIAA Paper No. 80-0214.
99. Sheu, T. S., Kuo, J. S., Ding, P. P., and Chen, P. H. (2004). *Microsyst. Technol.* **10**(4), 315–322.
100. Wu, H. Y. and Cheng, P. (2003). *Int. J. Heat Mass Transf.* **46**, 2519–2525.
101. Ayyaswamy, P. S., Catton, I., and Edwards, D. K. (1974). *ASME J. Appl. Mech.* **41**(2), 248–265.
102. Murkami, M., Ogushi, T., Sakurai in, Y., Masumoto, H., Furukawa, M., and Imai, R. (1987). Heat pipe heat sink. *In* "Proceedings of the 6th International Heat Pipe Conference, May 25–28, 1987." Vol. 2, pp. 537–542. Grenoble, France.
103. Sato, K., Susa, Y., Tanaka, S., Kimura, Y., and Sotani, J. (1991). "Micro Heat Pipe." Yokohama Lab., Furukawa Electr. Co., Ltd., Yokohama, Japan. *Furukawa Denko Jiho* **88**, 29–34, CODEN: FKDJAX ISSN: 0429–9159.
104. Moon, S. H., Hwang, G., Ko, S. C., and Kim, Y. T. (2004). Experimental study on the thermal performance of micro-heat pipe with cross-section of polygon, *Microelectron. Reliab.* **44**, 315–321.
105. Zhou, J., Yao, Z., and Zhu, J. (1993). Experimental investigation of the application characters of micro heat pipe. *In* "Advances in Heat Pipe Science and Technology, Proceedings of the International Heat Pipe Conference – 8th, Beijing, September 14–18, 1992" (T. Ma, ed.), pp. 421–424. International Academic Publishers, Beijing, People's Republic of China, CODEN: 61QQA7.
106. Zhang, J., Wang, C., Yang, X., and Zhou, Z. (1993). Experimental investigation of the heat transfer characteristics of the micro heat pipes. *In* "Advances in Heat Pipe Science and Technology, Proceedings of the 8th International Heat Pipe Conference, Beijing, September 14–18, 1992" (T. Ma, ed.), pp. 416–420. International Academic Publishers, Beijing, People's Republic of China, CODEN: 61QQA7 Conference.
107. Alario, J., Haslett, R., and Kosson, R. (1981). The Monogroove High Performance Heat Pipe, AIAA paper No. 81-1156, June.
108. Babin, B. R., Peterson, G. P., and Wu, D. (1990). *J. Heat Transf.* **112**, 595–601.
109. Beam, J. E. and Mahafekey, E. T. (1986). Heat Transfer Visualization in the double wall artery heat pipe. *In* "AIAA/ASME Thermophysics and Heat Transfer Conference, June 2–4, 1986." Boston, MA, USA.
110. Stalmach, D. D., Oren, J. A., and Cox, R. L. (1984). "Systems Evaluation of Thermal Bus Concepts." 2–53200/2r-53030, February, Vought Crop. Dallas, TX, USA.
111. Fleischman, G. L. (1982). "Osmotic Heat Pipe." Air Force Wright Aeronautical Lab., TR-81-31332, November, Wright-Patterson AFB, Dayton, OH, USA.
112. Narasaki, T. (1978). The characteristics of bimorph vibrator pump. *In* "Proceedings of the SAE Energy Conservation Engineering Conference, August 1978." pp. 2005–2010. San Franscisco, CA, USA.
113. Peterson, G. P. (1986). *J. Spacecr. Rockets* **24**(1), 7–14.
114. Hanford, A. J. and Ewert, M. K. (1996). "Advanced Active Thermal Control Systems Architecture Study." NASA TM 104822, October, Houston, TX, USA.
115. Jones, T. B. (1974). *Mech. Eng.* **96**(1), 27–32.
116. Jones, T. B. and Perry, M. P. (1974). *J. Appl. Phys.* **45**(5), pp. 2129–2132.

117. Loehrke, R. I. and Debs, R. J. (1975). Measurements of the Performance of an Electrohydrodynamic Heat Pipe, AIAA Paper 75-659, May.
118. Bologa, M. K. and Savin, I. K. (1990). Electrohydrodynamic heat pipes. *In* "Proceedings of the 7th International Heat Pipe Conference (Minsk)." pp. 549–562. Begell House, New York, NY, USA.
119. Babin, B. R., Peterson, G. P., and Seyed-Yagoobi, J. (1993). *J. Thermophys. Heat Transf.* **7**(2), 340–345.
120. Sato, M., Nishida, S., and Noto, F. (1992). Study on electrohydrodynamical heat pipe. "Proceedings of the ASME JSES KSES International Solar Energy Conference, Part 1 of 2." pp. 155–160. American Society of Mechanical Engineers, New York, NY, USA.
121. Melcher, J. R. (1974). *IEEE Trans. Educ.* **E-17**(2), 100–110.
122. Bryan, J. E. and Seyed-Yagoobi, J. (1997). *J. Thermophys. Heat Transf.* **11**(3), 454–460.
123. Yu, Z. Q., Hallinan, K., Bhagat, W., and Kashani, R. (2002). *J. Thermophys. Heat Transf.* **16**(2), 180–186.
124. Yu, Z. Q., Hallinan, K., Bhagat, W., and Kashani, R. (2000). Experimental and analytical studies of the maximum heat transport capacity in electrohydrodynamically enhanced micro heat pipes. *In* "Proceedings of the 34th National Heat Transfer Conference, Pittsburgh, PA, United States, August 20–22, 2000" (S. C. Yao and A. Jones, eds.), Vol. 1, pp. 835–864. American Society of Mechanical Engineers, New York, NY, USA, CODEN: 69AUFY.
125. Yu, Z., Kelvin, P. H., and Kashani, R. (2003). *Exp. Thermal Fluid Sci.* **27**, 867–875.
126. Suman, B. (2006). *Int. J. Heat Mass Transf.* **49**, 3957.
127. Wu, D., Peterson, G. P., Chang, W. S. (1991). *J. Thermophys. Heat Transf.* **5**(4), 539–544.
128. Sobhan, C. B., Xiaoyang, H., and Yu, L. C. (2000). *J. Thermophys. Heat Transf.* **14**(2), 161–169.
129. Incropera, F. P. and DeWitt, D. P. (2000). "Fundamentals of Heat and Mass Transfer." 4th edn. John Wiley & Sons Inc. New York, NY, USA.
130. Ochterbeck, J. M. (2003). *In* "Heat Pipes in Heat Transfer Handbook" (A. Bejan and A. D. Kraus, eds.), John Wiley & Sons Inc, New York, NY, USA.
131. Duncan, A. B. and Peterson, G. P. (1994). *ASME HTD* **278**, 1–10, CODEN: ASMHD8 ISSN:0272-5673. CAN 124:32603 AN 1995:970183 CAPLUS.
132. Duncan, A. B. and Peterson, G. P. (1995). Charge optimization for triangular shaped etched micro heat pipe. *AIAA J. Thermophys. Heat Transf.* **9**(2), 365–367.
133. Lee, M., Wong, M., and Zohar, Y. (2003). *J. Micromech. Microeng.* **13**, 58–64.
134. Suman, B. (2007). Effects of a surface-tension gradient on the performance of a microgrooved heat pipe: an analytical study. *Microfluidics and Nanofluidics.* **5**, 655–667.
135. Chen, Y. P. and Cheng, P. (2003). *Int. Commun. Heat Mass Transf.* **30**(1), 1–11, January.
136. White, F. M. (1991). "Viscous Fluid Flow." 2nd edn. McGraw-Hill, New York, NY, USA.
137. Ma, H. B. and Peterson, G. P. (1997). *Microscale Thermophys. Eng.* **1**, 253.
138. Gokhale, A. M. and Drury, W. J. (1990). *Metallurgical Trans. A* **21A**, 1201.
139. Kang, S. W., Chen, J. S., and Hung, J. Y. (1998). *Int. J. Mech. Tools Manufacture* **38**, 663.
140. Ma, H. B. and Peterson, G. P. (1997). *Microscale Thermophys. Eng.* **1**(3), 253–265.
141. Suh, J. S., Greif, R., and Grigoropoulos, C. (2001). *Int. J. Heat Mass Transf.* **44**, 3103–3109.
142. Gerner, F. M. (1990). "Micro Heat Pipes." AFSOR Final Report No. S-210-10MG-066, Wright-Patterson AFB, Dayton, OH, USA.
143. Mallik, A. K., Peterson, G. P., and Weichold, W. (1991). Construction processes for vapor deposited micro heat pipes. *In* "10th Symposium on Electronic Materials Processing and Characteristics, June 3–4, 1991." Richardson, TX, USA.
144. Mallik, A. K., Peterson, G. P., and Weichold, M. H. (1995). *J. Microelectromec. Syst.* **4**(3), 119–31, CODEN: JMIYET ISSN: 1057-7157. CAN 124:19121 AN 1995:910084 CAPLUS.

145. Kang, S. W., Tsai, S. H., and Chen, H. C. (2002). *Appl. Thermal Eng.* **22**, 1559–1568.
146. Berre, M. L., Launay, S., Sartre, V., and Lallemand, M. (2003). *J. Micromech. Microeng.* **13**, 436–441.
147. Moon, S. H., Yun, H. G., Ko, S. C. Hwang, G., Choy, T. G., Jun, C. H., and Kim, Y. T. (2003). Micro Heat Pipe with Polygonal Cross-Section Manufactured Via Extrusion or Drawing, United States Patent Application 20040112572.
148. Lee, M.; Wong, M., and Zohar, Y. (2003). *J. Microelectromech. Syst.* **12**(2), 138–146, CODEN: JMIYET ISSN: 1057-7157.
149. Lee, M., Wong, M., and Zohar, Y. (2002). Design, fabrication and characterization of an integrated micro heat pipe. *In* "15th IEEE International Conference on Micro Electro Mechanical Systems, Technical Digest, Las Vegas, NV, USA, January 20–24, 2002." pp. 85–88. Institute of Electrical and Electronics Engineers, New York, NY, USA, CODEN: 69DOEU Conference written in English. CAN 138:155577 AN 2003:95112 CAPLUS.
150. Kang, S.-W., Tsai, S.-H., and Chen, H.-C. (2002). *Appl. Thermal Eng.* **22**(14), 1559–1568, CODEN: ATENFT ISSN: 1359-4311.
151. Heine, D. and Groll, M. (1984). Compatibility of organic fluids with commercial structure materials for use in heat pipes. *In* "Proceedings of the 5th International Heat Pipe Conference." pp. 170–174. Tsukuba, Japan.
152. Sugumar, D. and Tio, K. K. (2003). Heat transport limitation of a triangular micro heat pipe. *In* "1st International Conference on Microchannels and Minichannels, Rochester, NY, USA, April 24–25, 2003" (S. G. Kandlikar, ed.), pp. 739–746. American Society of Mechanical Engineers, New York, NY, USA, CODEN: 69FLEV.
153. Ma, H. B. and Peterson, G. P. (1996). The capillary limit of a micro heat pipe. *In* "Heat Transfer Science and Technology 1996 [4th International Symposium on Heat Transfer], Beijing, October 7–11, 1996" (W. Buxuan, ed.), pp. 341–347. Higher Education Press, Beijing, People's Republic of China, CODEN: 65CDAS.
154. Beavers, G. and Josef, D. D. (1967). *J. Fluid Mech.* **30**, 197–207.
155. Derjaguin, B. V., Nerpin, S. V., and Churaev, N. V. (1965). *Bull. Rilem* **29**, 93–98.
156. Potash, M. Jr. and Wayner, P. C., Jr. (1972). *Int. J. Heat Mass Transf.* **15**, 1851–1863.
157. Raif, R. J., Wayner, P. C. Jr. (1973). *Int. J. Heat Mass Transf.* **16**, 1919–1929.
158. Kamotani, Y. (1978). Evaporator film coefficients of grooved heat pipes. *In* "Proceedings of the Third International Heat Pipe Conference, American Institute of Aeronautics and Astronautics." pp. 128–130.
159. Renk, F. J. and Wayner, P. C., Jr. (1979). *J. Heat Transf.* **101**, 59–62.
160. Wayner, P. C., Jr., Kao, Y. K., and LaCroix, L. V. (1976). *Int. J. Heat Mass Transf.* **19**, 487–492.
161. Holm, F. W. and Goplen, S. P. (1979). *J. Heat Transf.* **101**, 543–547.
162. Mirzamoghadam, A. V. and Catton, I. (1988). *J. Heat Transf.* **110**, 201–207.
163. Stephan, P. and Busse, C. A. (1992). *Int. J. Heat Mass Transf.* **35**, 383–391.
164. Swanson, L. W. and Herdt, G. C. (1992). *J. Heat Transf.* **114**, 434–441.
165. Brown, J. R. and Chang, W. S. (1993). Heat transfer from stable evaporating thin films in the neighborhood of a contact line. *In* "National Heat Transfer Conference, Atlanta, Georgia, August 8–11, 1993".
166. DasGupta, S., Kim, I. Y., Wayner, P. C., Jr. (1994). *J. Heat Transf.* **116**, 1007–1015.
167. Swanson, L. W. and Peterson, G. P. (1995). *J. Heat Transf.* **115**, 195–201.
168. Wayner, P. C., Jr. (1999). *AIChE J.* **45**, 2055–2068.
169. Sharp, R. R. (1964). "The Nature of Liquid Film Evaporation During Nucleate Boiling." NASA-TN-D1997, Washington, DC, USA.
170. Jawurek, H. H. (1969). *Int. J. Heat Mass Transf.* **12**, 843–848.

171. Voutsinos, C. M., Judd, R. L. (1975). *J. Heat Transf.* **97**, 88–93.
172. Cook, R., Tung, C. Y., and Wayner, P. C., Jr. (1981). *J. Heat Transf.* **103**, 325–330.
173. Wayner, P. C., Jr., Tung, C. Y., Tirumala, M., and Yang, J. H. (1985). *J. Heat Transf.* **107**, 182–189.
174. Chebaro, H. C., Hallinan, K. P., Kim, S. F., and Chang, W. S. (1993). Evaporation from a porous wick heat pipe for non sothermal interfacial conditions. *In* "National Heat Transfer Conference, Atlanta, GA, USA, August 8–11, 1993".
175. Kim, I. Y. and Wayner, P. C., Jr. (1996). *J. Thermophys. Heat Transf.* **10**, 320–325.
176. Karthikeyan, M., Huang, J., Pawsky, J. L., and Wayner, P. C., Jr. (1998). *J. Heat Transf.* **120**, 166–173.
177. Kihm, K. D. and Pratt, D. M. (1999). Contour mapping of thin liquid film thickness using Fizeau interferometer. *In* "Proceedings of the 33rd National Heat Transfer Conference, Albuquerque, NM, USA, August 15–17, 1999." NHTC99-224.
178. Wang, Y.-X., Plawsky, J. L., and Wayner, P. C., Jr. (2000). Heat and mass transfer in a vertical constrained vapor bubble heat exchanger using ethanol. *In* "Proceedings of the 34th National Heat Transfer conference, Pittsburgh, PA, USA, August 20–22, 2000." NHTC2000-12201.
179. Wang, Y.-X., Plawsky, J. L., and Wayner, P. C., Jr. (2001). *Microscale Thermophys. Eng.* **1**, 55–69.
180. Jiang, P. and Carey, V. P. (2001). *ASME HTD* **369**(3), 1–13. CODEN: ASMHD8 ISSN:0272-5673. CAN 136:218849 AN 2002:121500 CAPLUS.
181. Basu, S., Joel, L. P., and Wayner, P. C., Jr. (2004). *Ann. NY Acad. Sci.* **1027**, 317–329.
182. Zhang, L. A., Wang, E. N., Goodson, K. E., and Kenny, T. W. (2005). *Int. J. Heat Mass Transf.* **48**(8), 1572–1582, April.
183. Hetsroni, G., Klein, D., Mosyak, A., Segal, Z., and Pogrebnyak, E. (2004). *Microscale Thermophys. Eng.* **8**(4), 403–421, October–December.
184. Peng, X. F., Peterson, G. P., and Wang, B. X. (1996). *Int. J. Heat Mass Transf.* **39**(6), 1257–1264, April.
185. Coleman, J. W. and Garimella, S. (1999). *Int. J. Heat Mass Transf.* **42**(15), 2869–2881, August.
186. Chen, Y. P. and Cheng, P. (2005). *Int. Commun. Heat Mass Transf.* **32**(1–2), 175–183, January.
187. Tien, C. L. and Chung, K. S. (1979). *AIAA J.* **17**(6), 643–646.
188. Kim, B. H. and Peterson, G. P. (1995). *Int. J. Heat Mass Transf.* **38**(8), 1427–1442.
189. DasGupta, S., Schonberg, J. A., Kim, I. Y., and Wayner, P. C., Jr. (1993). *J. Colloid Interface Sci.* **157**, 332–342.
190. DasGupta, S., Schonberg, J. A., and Wayner, P. C., Jr. (1993). *J. Heat Transf.* **115**, 201–208.
191. Gee, M. L., Hearly, T. W., and White, L. R. (1989). *J. Colloid Interface Sci.* **131**(1), 18–23.
192. Gokhale, S., Plawsky, J. L., Wayner, P. C. Jr., and DasGupta, S. (2004). *Phys. Fluids* **16**(6), 1942–1955.
193. Moosman, S. and Homsy S. M. (1980). *J. Colloid Interface Sci.* **73**, 212–223.
194. Troung, J. G. and Wayner, P. C., Jr. (1987). *J. Chem. Phys.* **87**, 4180–4188.
195. Peles, Y. P. and Haber, S. (2000). *Int. J. Multiphase Flow* **26**, 1095–1115.
196. Zhao, T. S. and Liao, Q. (2002). *Int. J. Heat Mass Transf.* **45**(13), 2829–2842.
197. Du, X. Z. and Zhao. T. S. (2003). *Int. J. Heat Mass Transf.* **46**(24), 4669–4679, November.
198. Daniel, S., Chaudhury, M. K., and Chen, J. C. (2001). *Science* **291**, 633.
199. Brennan, P. J. and Kroliczek, E. J. (1979). "Heat Pipe Design Handbook." B&K Engineering, Towson, MD, USA.
200. Cotter, T. P. (1964). "Theory of Heat Pipes, Los Alamos National Laboratory." Report No. LA-3246-MS, The University of California, Los Alamos, NM, USA.

201. Chisholm, D. (1971). "The Heat Pipe." Mills and Boon, London, England.
202. Tien, C. L. (1975). *Annu. Rev. Fluid Mech.* **2**, 167–186.
203. Feldman, K. T. (1976). "The Heat Pipe: Theory, Design and Applications, Technology Application Center." University of New Mexico, Albuquerque, NM, USA.
204. Faghri, A. (1989). *J. Heat Transf.* **111**, 851–857.
205. Riabov, V. V. and Provotorov, V. P. *J. Themophys. Heat Transf.* **9**(2), 363–365.
206. Tien, C. L., Majumdar, A., and Gerner, F. M. (1998). "Microscale Energy Transport." Taylor & Francis, Washington, DC, USA.
207. Chen, G. (2005). "Nanoscale Energy Transfer and Conversion." Oxford University Press.
208. Goodson, K. E. and Ju, Y. S. (1999). *Annu. Rev. Mater. Sci.* **29**, 261.
209. Chen, G. (1996). *Annu. Rev. Heat Trans.* **7**, 69.
210. Chen, G. (2001). *In* "Editor, Recent Trends in Thermoelectric Materials Research III (Semiconductors and Semimetals Vol. 71)" (T. M. Tritt, ed.), p. 203. Academic Press, San Diego, CA, USA.
211. Chen, G. and Shakouri, A. (2002). *Trans. ASME J. Heat Transf.* **124**, 242.
212. DiMatteo, R. S., Greiff, P., Finberg, S. L., Young-Waithe, K., Choy, H. K.H., Masaki, M. M., and Fonstad, C. G. (2001). *Appl. Phys. Lett.* **79**, 1894.
213. Li, Q. Tsay, Y. N., Suenaga, M., Wirth, G., Gu, G. D., and, Koshizuka, N. *Appl. Phys. Lett.* **74**, 1323–1325.
214. Flik, M. I., Choi, B. I., and Goodson, K. E. (1992). *ASME J. Heat Trans.* **114**, 667.
215. Tien, C. L. and Chen, G. (1994). *ASME J. Heat Trans.* **116**, 799.
216. Chen, G. (2001). "Semiconductors and Semimetals." Vol. 71, p. 203. Academic Press, San Diego.
217. Cahill, D. G., Ford, W. K., Goodson, K. E., Mahan, G. D., Majumdar, A., Maris, H. J., and Merlin, R. (2003). *J. Appl. Phys.* **93**, 743.
218. Domoto, G. A., Boehm, R. F., and Tien, C. L. (1972). *ASME J. Heat Trans.* **92**, 412.
219. Polder, D. and Von Hove, M. (1971). *Phys. Rev. B* **4**, 3303.
220. Dresselhaus, M. S. (2003). Nanostructures and energy conservation. *In* "Proceedings of 2003 Rohsenow Symposium on Future Trends of Heat Transfer, MIT, Cambridge, MA, May 16, 2003".
221. Kim, P., Shi, L., Majumdar, A., and McEuen, P. L. (2001). *Phys. Rev. Lett.* **87**, 215502.
222. Zheng, L., Plawsky, J. L., Wayner, P. C. Jr., and DasGupta, S. (2004). *J. Heat Transf.* **126**, 169.

A Review of Heat Transfer in Nanofluids

SARIT K. DAS[1,*] and STEPHEN U. S. CHOI[2]

[1]*Department of Mechanical Engineering, Heat Transfer and Thermal Power Laboratory, Indian Institute of Technology Madras, Chennai, Tamil Nadu 600 036, India*
[2]*Department of Mechanical and Industrial Engineering (MC 251), Nanofluids Laboratory, University of Illinois at Chicago, Chicago, Illinois 60607-7022*

I. Introduction

The later part of the 20th century belongs to the semiconductor revolution. Keeping other developments such as space technology, nuclear reactor technology, and biotechnology in mind, the development that has accelerated all other developments is definitely the innovations in electronic hardware with tremendous implications in computer science, communication, measurement, and control. The core of this revolution lies in miniaturization, packing millions of devices inside vanishing small dimensions. In 1959 celebrated physicist Richard Feynman presented [1] the idea of micro machines at the annual meeting of the American Physical Society. Today it is worth looking back at those predictions to find that reality has taken over imagination. However, this journey to the present ultra thin devices is not likely to continue unabated. Designers of electronic and computing devices are already feeling the bottleneck that they have reached. Surprisingly, the bottleneck is not electronic in nature but thermal. The movement toward the smaller devices with increasing speed of operation brings about ever-increasing heat flux. Few watts generated within a device of few nanometers thickness can give heat flux next to that of the sun. As the premise "bigger is better" gave way to "smaller is better" leading to the threshold of the new "nano revolution," the problems with thermal management of devices are going to multiply. Thus, one of the major challenges of the "nano-age" as Rohrer [2] puts it is the thermal management of shrinking devices.

The other important area that has experienced similar increasing problem in thermal management is the area of optical devices. Lasers, high-power X-rays, and optical fibers are integral parts of today's computation,

---

*Current address: Department of Mechanical Engineering, Massachusetts Institute of Technology (MIT), 77 Massachusetts Avenue, Room 3-354A, Cambridge, MA 02139

scientific measurement, material processing, medicine, material synthesis, and communication devices. The increasing power of these devices with decreasing size calls for innovative cooling technology. Heat transfer in microscale is an area of research by itself that has been adequately reviewed by Duncan and Peterson [3] and Majumdar *et al.* [4]. All these reviews indicate that the conventional extended surface and microchannel technology [5] are inadequate for the new generation of semiconductor and optoelectronic devices. Choi *et al.* [6] has shown that power densities of $2,000\,W\,cm^{-2}$ can be managed by microchannel heat exchangers using subcooled liquid nitrogen. The studies on microchannel boiling such as those by Kandlikar [7,8], Kandlikar and Grande [9], Bergles *et al.* [10], and Thome *et al.* [11] are on the rise indicating the limitation of convectional single phase cooling for micro-sized devices. The advent of nanoelectromechanical system (NEMS) and microelectromechanical system (MEMS) has only intensified this need asking for a paradigm shift in the cooling technology to keep pace with the ever-increasing speed of operation for electronic and optical devices.

However, the electronic devices are not the only area looking for enhanced cooling technology. Large devices such as transportation trucks and energy related equipment such as various types of fuel cells, supercritical boilers and nuclear reactor cores are also in need of more efficient cooling systems – higher in cooling capacity with decrease in size.

Thus, at both extremes of the cooling requirement – big or small – enhanced cooling technology is in tremendous need for the development of these technologies. This can be done in two ways: first, introducing new design of the cooling device such as microchannels, micro heat exchangers, micro jet, and impingement; and second, by enhancing the heat transfer capability of the fluid itself by producing engineered fluids. This chapter presents studies on this second route using suspensions of nanoparticles (typically less than 100 nm in size) in conventional fluids called nanofluids. Nanofluids differ from usual micro-sized suspensions and hence their uniqueness with respect to thermal properties can be the beginning point of discussion.

II. From Slurries to Nanofluid

The above challenges in cooling technology call for more efficient heat transfer fluids. In the development of energy-efficient heat transfer fluids, the thermal conductivity of the fluids plays a vital role. The most commonly used coolant, air, is almost insulating toward heat conduction, which is the reason for its inability to cool heat fluxes over 100 W cm^{-2} in electronic devices and as a result such systems will necessitate the use of liquid cooling. Recently, single-phase liquid cooling technologies such as microchannel heat sink and two-phase liquid cooling technologies such as heat pipes, thermosyphons, direct immersion cooling, and spray cooling for chip or package level cooling have emerged. Despite considerable research and development efforts on heat transfer enhancement, major improvements in liquid cooling capabilities have been constrained because traditional heat transfer fluids used in today's thermal management systems such as water, oils, and ethylene glycol have inherently poor thermal conductivities, orders of magnitude smaller compared to most solids. Thus, the usual techniques of heat transfer enhancement such as providing more extended surfaces, creating greater turbulence, or using active and passive devices for disrupting boundary layer appears to be "penny wise and pound foolish" because they do not recognize that a large extent of the problem lies in the fluid itself. Due to increasing cooling demand, a number of industries have a strong need, both at micro and macro levels, to develop advanced heat transfer fluids with significantly higher thermal conductivities than are presently available.

It is well known that at room temperature, metals in solid form have orders of magnitude higher thermal conductivities than those of fluids. For example, the thermal conductivity of copper at room temperature is about 700 times greater than that of water and about 3000 times greater than that of engine oil as shown in Table I. Therefore, the thermal

TABLE I

THERMAL CONDUCTIVITY OF VARIOUS MATERIALS

	Material	Thermal conductivity (W m^{-1} K^{-1})
Metallic	Silver	429
	Copper	401
	Aluminum	237
Nonmetallic	Diamond	3300
Solids	Carbon nanotubes	3000
	Silicon	148
	Alumina (Al$_2$O$_3$)	40
Metallic liquid	Sodium @ 644K	72.3
Nonmetallic	Water	0.613
Liquids	Ethylene glycol	0.253
	Engine oil	0.145

conductivities of fluids that contain suspended solid metallic particles could be expected to be significantly higher than those of conventional heat transfer fluids.

For more than 100 years, scientists and engineers have made great efforts to enhance the inherently poor thermal conductivity of liquids by adding solid particles in liquids. Numerous theoretical and experimental studies of the effective thermal conductivity of suspensions that contain solid particles have been conducted since Maxwell [12] presented a theoretical basis for predicting the effective conductivity of suspensions more than 100 years ago. This was followed by numerous theoretical and experimental studies such as by Hamilton and Crosser [13] and Wasp [14]. At the first sight, these models show encouraging promise with significant rise in fluid conductivity due to high conductivities of the solids. However, all of the studies on thermal conductivity of suspensions have been confined to millimeter- or micrometer-sized particles. This conventional approach has some major technical problems:

1. The conventional millimeter- or micrometer-sized particles sediment rapidly forming a layer on the surface. This fouling layer reduces the heat transfer effectiveness of the fluid.
2. If the fluid is circulated at a higher rate, sedimentation reduces but erosion of the heat transfer devices, pipelines, and pump power increases rapidly.
3. The large size of the particles in conventional suspensions does not work with the emerging "miniaturized" devices because they can clog the tiny channels of these devices.
4. The pressure drop in the fluid increases considerably due to increased viscosity.
5. Finally, we get only moderate conductivity enhancement based on particle concentration. That is, the more the particles, the more is the enhancement and the more are the problems indicated above.

Thus, although known for more than a century, the use of suspensions containing millimeter- or micrometer-sized particles was never considered as a serious candidate as a cooling fluid.

Modern nanotechnology has enabled the production of metallic or non-metallic nanoparticles with average crystallite sizes below 100 nm. The mechanical, optical, electrical, magnetic, and thermal properties of nanoparticles are superior to those of conventional bulk materials with coarse grain structures. Coupled with new studies on microchannel flow [15] related to heat transfer, this has given a new option of revisiting suspensions with nanoparticles. Recognizing an excellent opportunity to apply nanotechnology to thermal engineering, Choi [16] at Argonne National

Laboratory (ANL), Chicago, conceived the novel concept of nanofluids from the vision that it is possible to break down these century-old technical barriers by exploiting the unique properties of nanoparticles. They also coined the term "nanofluids". Nanofluids are a new class of nanotechnology-based heat transfer fluids engineered by dispersing nanometer-sized particles with typical length scales on the order of 1–100 nm (preferably, smaller than 10 nm in diameter) in traditional heat transfer fluids. It will not be out of context to indicate here that biologists and physicists have been using the term "nanofluid" for different types of particles such as DNA, RNA, proteins, or fluids contained in nanopores [17–19]. However, we will exclude them from this chapter.

Due to their large surface area, less particle momentum, and high mobility, nanoparticles emerged as suitable candidates for suspending in fluids. At the very first sight, these are expected to give higher effective conductivity, less sedimentation, less abrasion, and less clogging. This was the rationale behind trying nanofluids for heat transfer applications. On the conductivity enhancement side right from copper, one can even go up to multiwalled carbon nanotubes (MWNTs) which at room temperature has got 20,000 times greater conductivity than engine oil [20]. This made a preamble for trying suspensions with nanoparticles for heat transfer applications. The fact that only small volume fraction of the particles is required not only increases its stability but also reduces the increase in pumping loss due to increased viscosity. Nanofluids are also likely to reduce the size of heat transfer systems, giving smaller fluid inventory and less material of construction resulting in a substantial energy and cost savings. However, a serious concern remains about the issue of agglomeration as the particle sizes go down to nanoscale. Das *et al.* [21] have seen considerable amount of agglomeration of Al_2O_3 and CuO particles under atmospheric conditions. How well these agglomerates can be dispersed depends on the method of dispersion used. We will deal with this issue separately.

III. Nanofluids: Novel Features

The above observations gave impetus to begin research in nanofluids with an expectation that these fluids will play an important role in developing the next generation of cooling technology. This research needed to concentrate not only on the fundamentals of thermal transport and applications but also on synthesis, characterization, and fundamental physics aspects. Before going to these details, let us first have an overview of the novel features that have been observed in nanofluids. Interestingly, the first test with nanofluids gave more encouraging features [22] than it was thought to possess. The unique features observed were:

1. *High enhancement of thermal conductivity*: The most important feature observed in nanofluids was an abnormal raise in thermal conductivity much beyond any existing theory can predict.
2. *High stability*: Nanofluids have been reported to be stable over weeks and months with no or very small amount of stabilizing agents [23]. However, factors such as solution pH value and the IEP (isoelectric point) of the particles play important roles in preventing particle agglomeration.
3. *Small concentration, Newtonian behavior*: It was found that large enhancement of conductivity can be achieved with very small concentration of particles that keeps the fluid Newtonian in behavior. The rise in viscosity was nominal and hence pressure drop was found to increase only marginally.
4. *No erosion or clogging*: Due to the small size of the particles, they cannot impart large momentum on the solid surfaces; as a result no erosion was observed. Also no clogging in the nanofluid operation was observed primarily due to the small size of the particles.
5. *Particles size dependence*: Unlike microsuspensions, the enhancement of conductivity was found to depend on particle size. With decreasing size increase in enhancement was observed. The reason for this is needed to be investigated next.

The intention of listing the above observations was to present in brief the extent in which the projected potential has been realized. Here, we have not presented any detail or references of these observations, which subsequently will be made in the appropriate sections. The above features only indicate the reason for the stream of studies pouring in this area. Looking at the streams of studies, few interesting reviews have already come out in this area. Das *et al.* [24] as well as Wang and Mujumdar [25] have presented a critical review of the literature on thermal conductivity, convection, boiling studies, and theoretical investigations of heat transfer in nanofluids. Daungthongsuk and Wongwises [26] present a review on convective heat transfer using nanofluids. Gandhi [27] also gives a qualitative review of the experimental findings as well as different mechanisms in the heat transfer in nanofluids.

A review of an early literature is also presented by Trisaksri and Wongwises [28]. However, all these reviews are more of a qualitative in nature and many of them are on specific area of nanofluid research. Hence, there is a need to provide a comprehensive review from all the related disciplines to indicate the future direction of development in nanofluids which is attempted here.

IV. Synthesis of Nanoparticles

The review of nanofluids cannot be complete without a discussion on the synthesis or in more general terms production of nanoparticles. However, production of nanoparticle itself is too vast an area to review. An increasingly large amount of research is being done in this area primarily to use these particles to electrical and optical devises, catalysis, and biosciences. The techniques used in these applications are micro emulsion synthesis, chemical synthesis, mechanical attrition, precipitation, chemical vapor deposition, and LASER vapor deposition. Gleiter [29] provides a good overview of the synthesis methods. We will not go to the details of these techniques but will briefly mention only those techniques that are used for making particles for nanofluids.

The first materials tried for nanofluids were oxide particles primarily due to their ease of production and chemical stability in solution. Various investigators have produced Al_2O_3 and CuO nanopowder by inert gas condensation (IGC) process [22,23,30] giving particle sizes of 2–200 nm. In this process, the material is evaporated in a low-density inert gas, which is then condensed inside the gas, and the nanoparticles are transported and deposited by thermophoretic diffusion on a cold finger. Thereafter oxygen is introduced to produce oxides. The particles are scarped from the cold finger and dispersed in liquid. Usually the rate of production is low in IGC, but the recent development of gas condensation method [31] has been able to improve the yield. The major problem with this method is its tendency to form agglomerates and its unsuitability to produce pure metallic nanopowders.

The problem of agglomeration can be reduced to a good extent by using a direct evaporation condensation (DEC) method. In conventional dispersion methods, a dispersing chemical (acid, base, or surfactant) is used to prevent agglomeration and settling. This can also be done by condensing the vapor directly into a low vapor pressure fluid. This is called the vacuum evaporation onto a running oil substrate (VEROS) [32]. The DEC method is a modification of this method which has been adopted at ANL [33–35]. This method is very successful in producing pure Cu nanoparticles. Figure 1 shows the particle size (<10 nm) and dispersed nature of particles produced by this method. Even though this method has limitations of low vapor pressure fluids and oxidation of pure metals, it has excellent control over particle size and gives particles for stable nanofluids without any surfactant or electrostatic stabilizer.

The other method is the LASER vapor deposition technique used to produce SiC nanoparticles from SiH_4 and C_2H_4 [36]. In recent times, carbon nanotubes (CNTs) were used to produce nanofluids. The MWNT used for this purpose can be produced by chemical vapor deposition technique using xylene as carbon source and ferrocene as the catalyst [37].

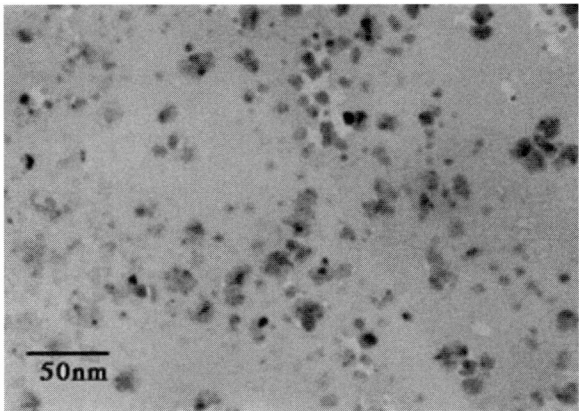

FIG. 1. Bright-field transmission electron microscope (TEM) micrograph of copper nanoparticles produced by direct evaporation in ethylene glycol [33].

Purely chemical synthesis method is the other option for producing nanoparticles that are known as metal clusters. The two methods by which nanofluids are made directly are described by Patel et al. [38]. The first one is the well-known citrate reduction method [39]. In this method, the chloroauric acid and silver nitrate are reduced using sodium citrate to produce gold and silver nanoparticles, respectively. The particles produced by this method were of 10–20 nm in size without any coating. However, this method leads some traces of acid in the suspension. It produces nanofluids with polar liquid like water for particles like gold and silver. This method can also be extended to nonnoble metal. The other method is the Brust [40] method where metal particles with monolayer of octadecanethiol is produced in powder form and then can be dispersed in solvent. The size of the particles is 4–8 nm. In the Brust method, ammonium bromide is added to chloroauric acid along with thiols, which on sonication produces gold nanoparticles. Figure 2 shows the bare particles and Fig. 3 the coated particles produced. The stability of these particles is more than the bare particles produced by citrate reduction method. The other option is the production of pure nanoparticles by ball and mill method as done by Xuan and Li [23]. However, the particle size obtained by them was rather large ($\sim$100 nm). Recently, Zhu et al. [41] has prepared nanofluids of metallic Cu nanoparticles dispersed in ethylene glycol by one-step chemical method in which $NaH_2PO_2 \cdot H_2O$ was added as reducing agent and microwave irradiation was used for heating. Microwave irradiation in comparison with conventional heating not only provided energy for heating but also accelerated the nucleation of Cu and depressed the straightforward growth of newborn Cu.

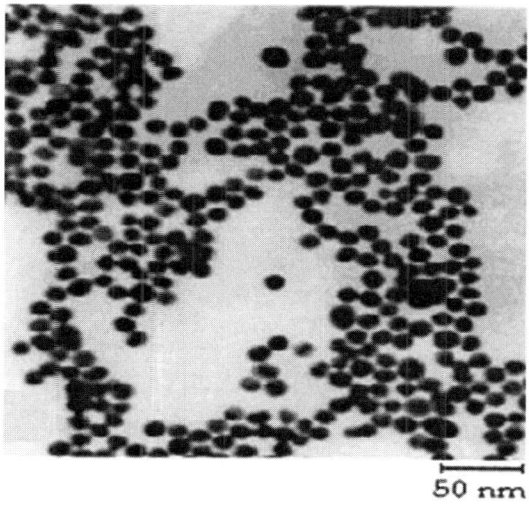

FIG. 2.   TEM photographs of the bare gold particles [38].

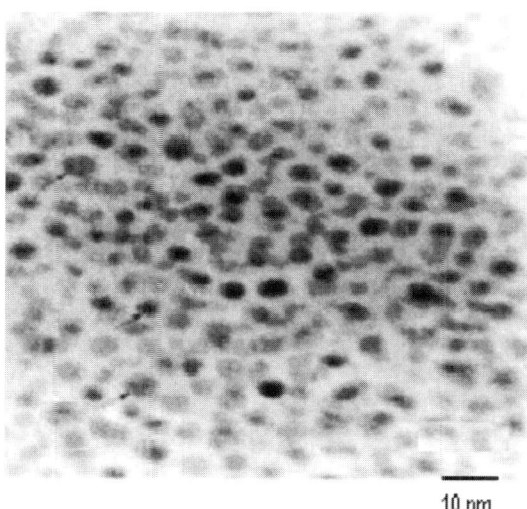

FIG. 3.   TEM photographs of the thiol-coated gold nanoparticles [38].

Thus a variety of physical, chemical, and LASER-based methods are available for the production of nanoparticles required for nanofluids. However, the task remains to characterize them and to disperse them in fluid.

V. Particle Characterization

A variety of techniques have been used to determine the particle size, shape, morphology, and size distribution of nanoparticles for nanofluid applications. Probably, TEM (transmission electron microscope) is the most widely used apparatus for this. Lee *et al.* [22] presented the TEM photograph of Al_2O_3 and CuO particles, which proved good dispersion with very few agglomerates. Das *et al.* [21] also showed that Al_2O_3 and CuO particles remain in micrometer-sized agglomerated form in air. This can be dispersed to nanometer size by suitable method. Figure 4 shows the agglomerated Al_2O_3 and Fig. 5 the dispersed particles of same Al_2O_3 as seen by TEM.

The studies at Argonne [22,33–35] used TEM to characterize the particles of size below 40 nm. Choi *et al.* [42] also used high-resolution scanning electron microscope (SEM) to characterize MWNT having mean diameter of ~25 nm and an aspect ratio of ~2000. They found the nanotubes to be fairly straight in configuration. This is shown in Fig. 6. The gold particles produced by Patel *et al.* [38] were also characterized by TEM as shown in Figs. 2 and 3. They were also characterized by usual techniques such as optical absorption (UV) spectroscopy [43].

Other simple methods were also used to characterize the nanoparticles. Xie *et al.* [44] used X-ray diffraction method to determine the crystalline phases of Al_2O_3 particles and Brunauer–Emmett–Teller method of nitrogen adsorption to measure the specific surface area. Das *et al.* [21] used the dynamic light scattering technique to find the particle sizes between 30 and

FIG. 4. TEM photograph of agglomerated Al_2O_3 nanoparticles [21].

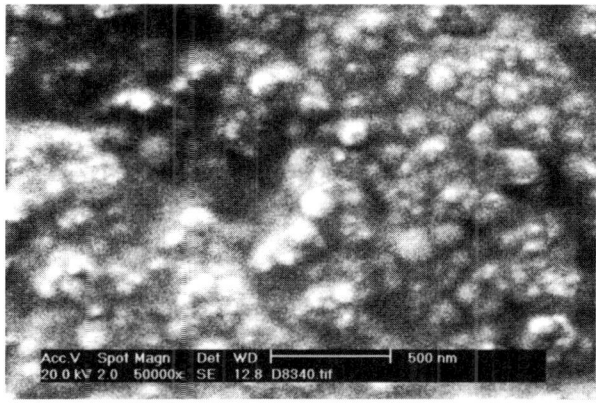

FIG. 5. TEM photograph of dispersed Al_2O_3 nanoparticles [21].

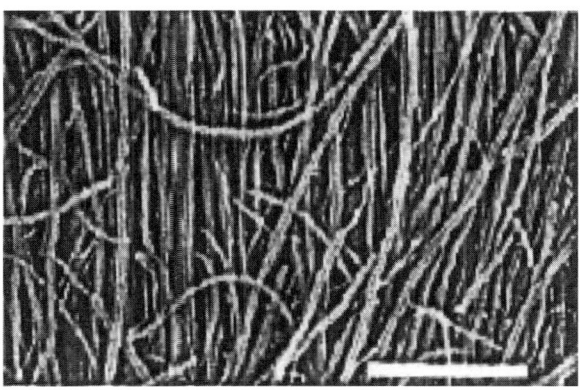

FIG. 6. Muti-walled nanotubes showing fairly straight configuration [42].

50 nm. In this method, the intensity of monochromatic light scattering fluctuates to diffusion coefficient of the particles and the particle size can be determined.

At this point, it can be said that it will also be interesting to characterize the particles with scanning tunneling microscope (STM) because these are the appropriate devices for metallic systems. Since the pure metallic particles show greater promise for nanofluid applications, STM may find their way in characterization in future. It must be mentioned here that all the characterizations done so far were for the particles alone and not for a nanofluid suspension. This may be an area of interest in future where particle characteristics in suspension will be studied.

VI. Preparation of Nanofluids

After production and characterization of nanoparticles, the most important step is to prepare the nanofluid. This is one of the most critical points because the nano-effect can only be observed only if the particles are well dispersed in the fluid. What differentiates the nanoparticles from the traditional microparticles is their ability to remain suspended while the microparticles easily settle down under gravity due to their weight. However, with decreasing size the particles may have a tendency to form agglomerates that can be of micrometer size and will eventually settle down in a way similar to microparticles. Thus, in order to have a stable nanofluid, the particles should be dispersed with no or very little agglomeration giving "true nano" behavior. This can be done by a variety of methods including electrical, physical, or chemical methods. However, the best way may be to produce them by single-step method, where instead of nanoparticles, the nanofluids are directly produced reducing the chance of agglomeration.

To understand the dependence of sedimentation ratio on different parameters, one can look at the explanation given by Choi et al. [45]. Let us consider a particle of spherical shape with radius r_p and density ρ_p. From Stokes–Einstein theory [46], the sedimentation rate of particles can be given by

$$v = \frac{2r_p^2 |\rho_p - \rho_f| g}{9\mu_f} \tag{1}$$

where ρ_f and μ_f are fluid density and fluid viscosity, respectively.

The equation indicates that sedimentation will decrease if the density difference is low, viscosity is high, and particles are small. Density difference can be reduced by the proper choice of material, but most of the metals, which are attractive from a conductivity point of view, have large density compared to water or oils. Increase of viscosity is obviously ruled out because of its adverse effect on heat transfer. Thus, the key to the issue of stability (smaller rate of sedimentation) is reduction of particle size. This essentially means breaking the agglomerates and dispersing the particles well into the liquid. The other factors that may play role are electrical charge, pressure of surfactants, mechanical agitation, and so on. A small amount of laurate salt was used by Xuan and Li [23] as stabilizing agent that improved the stability of pure copper particles drastically. Thus, nanofluids can be prepared by the physical or chemical dispersion techniques used for colloidal solutions [46,47].

A. Physical Dispersion Technique

The physical methods used so far are mechanical and ultrasonic dispersion. Mechanical method means breaking the agglomerates by high shear mixing. These are usually known as rotor–stator methods. Homogenizer, Kady mill, and colloid mill are some of the methods in their category. High impact grinding by small grinding material can also be used – attritor and ball-and-pebble mill are examples.

Compared to mechanical dispersion techniques, ultrasonic vibration is found to be better option for nanofluids. Das et al. [21] used ultrasonic vibration for 4 h to get excellent dispersion of nanoparticles. The ultrasonic vibration was created in water in a bath where the container containing primary solution of nanoparticles (agglomerates) was suspended. The ultrasonic sends an elastic wave giving mechanical and thermal interactions. The thermal interaction was evident from the elevated temperature after the dispersion process. It was found that the type of ultrasonic source (immersed type or bath type) did not make much of a difference but the time under vibration was found to be important. The time was reduced from 12 to 4 h by trial and error. However, no systematic study of the effect of ultrasonic frequency on dispersion was made. The other work in which multiple physical dispersion techniques were used came from Xie et al. [48] who dispersed SiC particles in ethylene glycol by ultrasonic and magnetic stirring.

B. Chemical Dispersion Method

Chemical methods are primarily aimed at disrupting long-range attractive Van der Waals force that is of the order of $k_b T$ (k_b = Boltzmann constant and T = absolute temperature). This can be done in a number of ways. Chemical dispersions can be done by electrostatic, steric dispersion, or functional group coating technique. Basically, electrostatic method is to charge the particles with similar charges and create the repulsive electrostatic forces which are just sufficient to oppose the long-range Van der Waals forces. These charges can be brought about by ionic species such as detergents or adding electrolytes [49–51]. This is same as stabilization of suspensions by pH control. However, these forces are quite sensitive to the pH value and at certain concentration electrolyte may even promote agglomeration and sedimentation. They may also change the thermophysical as well as chemical properties of the base fluid that may be undesirable.

Compared to this, steric stabilization prevents agglomeration by surfactants. Polymeric surfactants are mostly used. The long chain of the organic molecules creates steric repulsion. A number of chemicals are used as surfactants. They usually use an anchor molecule and a stabilizing molecule. Anchor molecules can be a variety of polymers such as polyacrylonitrile, polyvinyl chloride (PVC), and polydimethylsiloxane along with stabilizing molecules such as polylauryl methacrylate or polyvinyl methyl ether. For aqueous systems, a separate group of chemicals are suggested with anchors such as polystyrene, polyvinyl acetate, and stabilizer such as polyvinyl alcohol or polymethacrylic acid. Xuan and Li [23] used a combined physical and chemical dispersion method where addition of oleic acid of fairly large amount (up to 22% by weight or wt) was followed by 10 h of ultrasonic vibration for oil-based Cu nanofluids and smaller amount (7% by wt) of laurate salt addition followed by similar duration of ultrasonic vibration for water-based nanofluids. Of course, one must appreciate that the particles used by them were much larger ($\sim$100 nm), keeping in mind that the stabilization was quite satisfactory.

C. SINGLE-STEP METHODS

The other stabilizing technique of surface coating has got the advantage that they can give the single-step method for producing nanofluids as suspensions itself without the need of separate particle production. Surface coating is a method that is much simpler from a chemical point of view, but it also acts as a thermal barrier between the particle surface and the fluid that may impede heat transfer. Usually for chemical dispersion, surface coating or surface modification is widely accepted [52]. This can be done by adding various functional groups to the surface. While making gold- and silver-based nanofluids, Patel et al. [38] used octadecanethiol for forming a monolayer around gold or silver nanoparticles. This was done by the well-known Brust [40] method. Similarly as described earlier, the single-step citrate reduction method [39] was used for production of nanofluids with bare gold and silver particles which can be termed as electrolytic stabilization through the chemical synthesis process itself. The direct evaporation–condensation (DEC) method used by the ANL group also produced highly stable nanofluids [22] with and without thioglycolic acid. In recent times, one of maximum thermal conductivity enhancement was reported in nanofluids with CNTs [37] which were also dispersed in poly(α-olefin) oil using usual dispersion technique. The possibility of producing MWNT-based nanofluids by a single-step method is under investigation.

Chen et al. [53] suggested using electrohydrodynamic spraying system [54] for production of nanofluids. Even though it is a two-step method in the strict sense, the steps can be so coupled that it gives a continuous production of nanofluid. This is because of the fact that the rate of particle production by this method is quite high (10 billion particles per second). The particles are charged and hence do not coagulate, and the particles can directly be injected into a flowing fluid which can be called a pseudo single-step method. The novel one-step method by Zhu *et al.* [41] is also notable.

VII. Thermal Conductivity Enhancement in Nanofluids

The fact that generated primary interest in nanofluids is that thermal conductivity of nanofluids is much improved compared to usual suspensions. The observed enhancement of effective thermal conductivity over that of the base fluid is often few times for nanofluid compared to what would have been given by usual micrometer-sized suspensions. The base fluids used so far are water, ethylene glycol, transformer oil, toluene, and so on, keeping the applications of these fluids for cooling in mind. The nanoparticles used can be broadly divided into three groups – ceramic particles, pure metallic particles, and CNTs. They can also classify differently since different combinations of these particles and fluids give different nanofluids. However, we will classify them mainly by the type of particles.

A. Ceramic Nanofluids

Nanofluids containing ceramic particles were the first type of nanofluids investigated by the ANL group. The first major publication in this area by Lee *et al.* [22] presented measurement with fluids of Al_2O_3 and CuO in water and ethylene glycol. The measurement was made by the traditional transient hot-wire method (THWM). The same technique was also used by Patel *et al.* [38] to measure thermal conductivity of metallic nanofluids. The method is simple and accurate. Figure 7 shows the basic electrical circuit for the measuring apparatus. The method is based on the principle that if a thin wire is heated by passing electricity through it, the rate of rise of its temperature depends on the rate of conduction of heat to the fluid surrounding the wire. The lower the conductivity of the surrounding medium, the faster will be the temperature rise of the wire. Hence the thermal conductivity can be evaluated from the time–temperature curve of the wire. The detail of one such apparatus is given in the next section.

The result of Lee *et al.* [22] clearly indicated that the thermal conductivity enhancement of the Al_2O_3 and CuO nanofluids was high. They used volume functions of only 1–5%. The enhancement was higher with ethylene glycol as

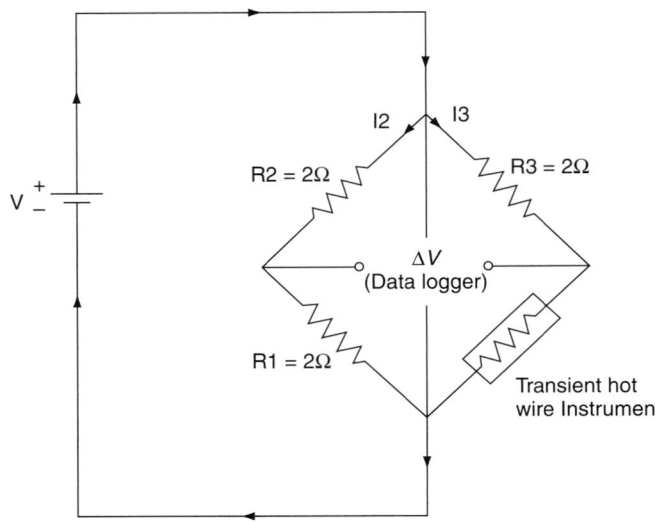

FIG. 7. Electrical circuit for transient hot-wire method [38].

base fluid. An enhancement of 20% was observed at 4% volume fraction of CuO. The enhancement with water as base fluid was lower but still substantial with 12% enhancement at 3.5% CuO and 10% enhancement with 4% Al_2O_3. The higher enhancement for CuO nanofluids could be their smaller size (23 nm) compared to Al_2O_3 particles (38 nm). Figure 8 shows the measurements of Lee et al. [22]. These results were high compared to the model for suspensions proposed by Maxwell [12] as early as in 1882. The theory of Maxwell [12] was later improved by Hamilton and Crosser [13] in 1962 to include the effect of particle shape. These models are essentially some kind of weighted average of solid and liquid conductivities.

The original Maxwell model reads as

$$\frac{k_{\text{eff}}}{k_{\text{f}}} = 1 + \frac{3(k_{\text{p}}/k_{\text{f}} - 1)\phi}{(k_{\text{p}}/k_{\text{f}} + 2) - (k_{\text{p}}/k_{\text{f}} - 1)\phi} \qquad (2)$$

Maxwell-Garnett [55] reads as

$$\frac{k_{\text{eff}}}{k_{\text{f}}} = \frac{(1-\phi)(k_{\text{p}} + 2k_{\text{f}}) + 3\phi k_{\text{p}}}{(1-\phi)(k_{\text{p}} + 2k_{\text{f}}) + 3\phi k_{\text{f}}} \qquad (3)$$

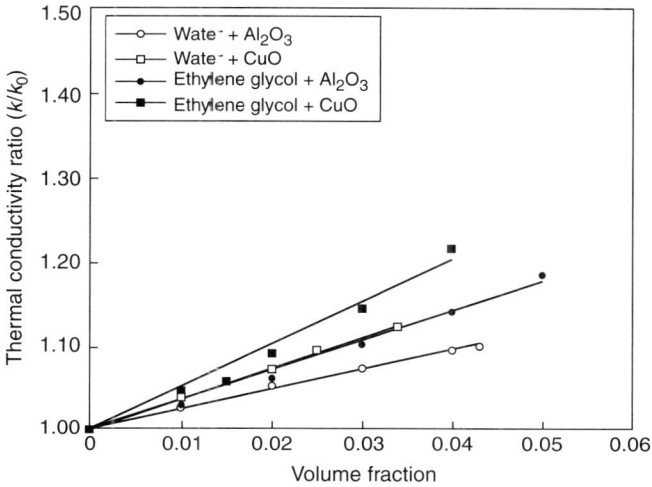

FIG. 8. Enhanced thermal conductivity of oxide nanofluid systems [22].

The Hamilton–Crosser [13] model reads as

$$\frac{k_{\text{eff}}}{k_{\text{f}}} = \frac{k_{\text{p}} + (n-1)k_{\text{f}} - (n-1)\phi(k_{\text{f}} - k_{\text{p}})}{k_{\text{p}} + (n-1)k_{\text{f}} + \phi(k_{\text{f}} - k_{\text{p}})} \quad (4)$$

where

k_{eff} = effective thermal conductivity of slurry;
k_{f} = thermal conductivity of liquid;
k_{p} = thermal conductivity of solid particles;
ϕ = volume fraction of nanoparticles; and
n = shape factor (for sphere = 3, for cylinder = 6).

It should be noted that both these correlations do not include particle size as a parameter. Figure 9 (a) and (b) shows the comparison of Lee *et al.*'s [22] measurement with the Hamilton–Crosser [13] model. It is interesting to note that Lee *et al.* [22] found that the Hamilton–Crosser [13] model was approximately able to predict the enhancement of Al_2O_3–water nanofluids (Fig. 9 (a)) but not the CuO–water nanofluids (Fig. 9 (b)). However it will be seen later that this match for prediction of Al_2O_3–water nanofluids was purely accidental due to temperature effect [21].

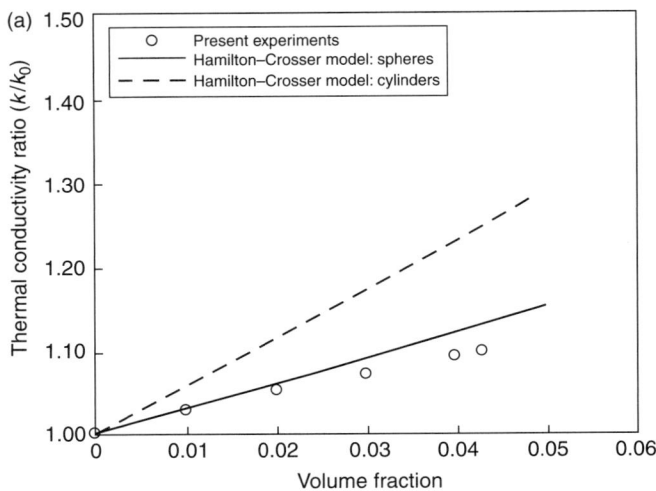

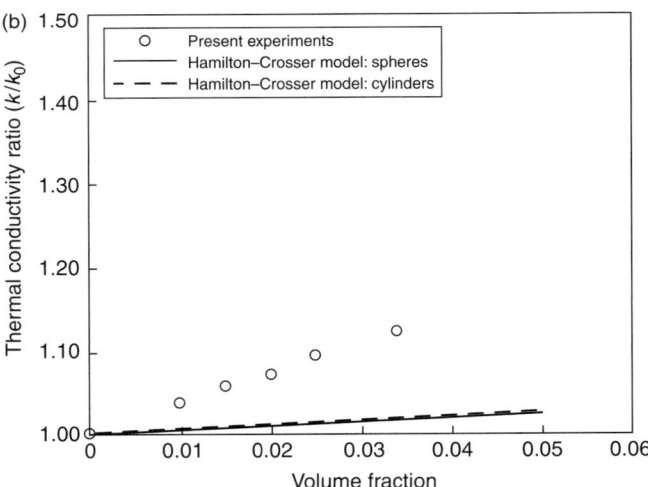

FIG. 9. (a) Comparison of increase of thermal conductivity ratio between Hamilton–Crosser model and experimental result with Al_2O_3/water nanofluids [22]. (b) Comparison of increase of thermal conductivity ratio between Hamilton–Crosser model and experimental result with CuO/water nanofluids [22].

A few higher-order models are also available which try to incorporate particle interactions by incorporating the higher-order terms such as these by Jeffrey [56]

$$\frac{k_{\text{eff}}}{k_{\text{f}}} = 1 + 3\beta\phi - \left(3\beta^2 + \frac{3\beta^2}{4} + \frac{9\beta^3}{16}\frac{\alpha+2}{2\alpha+3} + \cdots\right)\phi^2 \quad (5)$$

The model of Davis [57] reads as

$$\frac{k_{\text{eff}}}{k_{\text{f}}} = 1 + \frac{3\left[\frac{k_{\text{p}}}{k_{\text{f}}} - 1\right]}{\left(\frac{k_{\text{p}}}{k_{\text{f}}} + 2\right) - \left(\frac{k_{\text{p}}}{k_{\text{f}}} - 1\right)\phi}\left[\phi + f\left(\frac{k_{\text{p}}}{k_{\text{f}}}\right)\phi^2 + O(\phi^3)\right] \quad (6)$$

where

$$[\text{beta not in prev eqn}]\beta = \frac{\left(\frac{k_{\text{p}}}{k_{\text{f}}} - 1\right)}{\left(\frac{k_{\text{p}}}{k_{\text{f}}} + 2\right)} \quad (7)$$

and $f(10) = 2.5$

$f(8) = 0.5$

or the model of Lu and Lin [58]

$$\frac{k_{\text{eff}}}{k_{\text{m}}} = 1 + \frac{k_{\text{p}}}{k_{\text{f}}}\phi + b\phi^2 \quad (8)$$

Wang et al. [25,30] also measured the thermal conductivity of CuO and Al$_2$O$_3$–water nanofluids, but their particle size was smaller (23 nm for CuO and 28 nm for Al$_2$O$_3$). They also measured the nanofluids with ethylene glycol and engine oil (Pennzoil 10W-30) as the base fluids. The measurement showed a clear effect of particle size and the method of dispersion. These parameters will assume greater significance later on as we will see subsequently. Xie et al. [44] measured aqueous Al$_2$O$_3$ nanofluid but for even reduced particle size. Their range of particle size was 1.2–302 nm. They also observed effect of particle size and effect of pH value of the solution base. Thus, it has been generally found that oxide ceramic particles which themselves do not have very high thermal conductivity can enhance the thermal conductivity of fluids in nano suspensions. The main reason for many studies on oxide particle-based nanofluids is the availability of oxide nanoparticles and the relative ease with which stable suspension can be made with them. A number of companies today have started producing nanoparticles of oxide ceramics commercially. However, it has to be kept in mind that the investigators need to characterize them. Das *et al.* [21] found that the particle size distribution of Al_2O_3 nanoparticles is quite different from that suggested by the manufacturer. This affects the thermal conductivity values severely. Measurement for nonoxide ceramic (SiC particles with ethylene glycol) nanofluid was made by Xie *et al.* [48]. For particle sizes between 26 and 600 nm, they observed that at lower particle size the Hamilton–Crosser [13] model underpredicts the thermal conductivity while at higher size it predicts well.

Murshed *et al.* [59] measured the thermal conductivity of aqueous solution of TiO_2 nanoparticles of spherical as well as cylindrical shape. The 15-nm-sized spherical particles showed slightly less enhancement than that of 10×40 nm-sized rods that showed an enhancement of 33% for a volume fraction of 5%. However, the enhancements were far more than that the predictions of Hamilton–Crosser model. Another feature brought out in this work was the nonlinear dependence of enhancement in thermal conductivity on particle concentration at lower volume fractions. Zhua *et al.* [60] measured the thermal conductivities of Fe_3O_4 aqueous nanofluids and found that the enhancement is quite high (40% enhancement for 5% volume fraction of Fe_3O_4 nanoparticles of 9.5 nm size) and it has a strong nonlinear effect of volume fraction, with relatively higher enhancement at lower volume fraction. Hwang *et al.* [61] have measured the thermal conductivities of TiO_2 (25 nm), Al_2O_3 (48 nm), Fe (10 nm), and WO_3 (38 nm) nanofluids. TiO_2 in water nanofluid gave an enhancement of about 15% for 1% volume fraction, whereas alumina gave 4% enhancement for the same concentration in water. Fe showed an enhancement of 18% for 0.55% volume fraction in ethylene glycol, whereas WO_3 showed 13% enhancement for 0.3% volume fraction in ethylene glycol. In general, the enhancements were found to be high. Kim *et al.* [62] measured the thermal conductivity of alumina, ZnO, and TiO_2 particle suspensions in water and ethylene glycol. They analyzed particle size effect using ZnO (10, 30, and 60 nm) and TiO_2 (10, 35, and 70 nm) nanoparticles and found that there is not a very strong dependence of thermal conductivity on it.

B. METALLIC NANOFLUIDS

Although the ceramic nanofluids demonstrated the potential of nanofluid, a big step forward was the emergence of metallic particles based nanofluids. Xuan and Li [23] were the first to try copper particle-based nanofluids of transformer oil. Their enhancement was comparable with oxide nanofluids

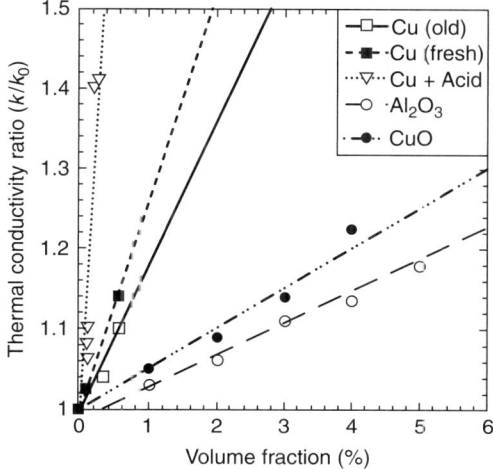

FIG. 10. Thermal conductivity enhancement for various nanofluids [33].

mainly due to the fact that they used a much larger (~100 nm) particle size due to limitation of their method of synthesis and they reported 55% enhancement for 5% volume fraction. It should also be kept in mind that due to the ball milling technique used by them, they had a wide size distribution of particles.

However, the real breakthrough was from the ANL group reporting a 40% enhancement of conductivity with only 0.3% copper particles of 10 nm size [33]. This clearly shows the particle size effect and the potential of nanofluids with smaller particle sizes. They were stabilized with thioglycolic acid. Figure 10 shows the measured values of thermal conductivity measurement for Cu–ethylene glycol nanofluids. They also observed significant difference between freshly prepared and old samples of nanofluids which can be attributed to oxidation of the particles. The figure also contains the values for ceramic nanofluids to indicate that the leap occurs in the metal-based nanofluids. This was probably the first one to indicate that the nanofluids can give significant progress in cooling technology. The study also indicated a mild "aging effect" with freshly prepared nanofluids showing higher conductivity.

In another very interesting study, Patel et al. [38] used pure metallic gold and silver for the first time to prepare nanofluids. They also used transient hot wire method for measuring thermal conductivity. Their test cell is shown in Fig. 11, which contains an insulating cylindrical container with a platinum wire at the axis. This wire acts as one arm of a Wheatstone bridge.

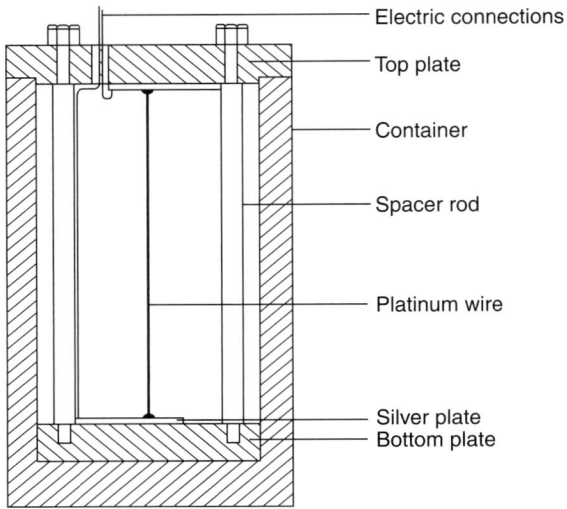

FIG. 11. Test cell for transient hot-wire method [38].

The platinum wire (100 μm diameter) is chosen because it can be used both as heater and thermometer due to its known temperature coefficient of resistance.

The thermal conductivity is evaluated from the heat conduction solution for a continuous, line source in a cylinder given by Carslaw and Jaeger [63] as

$$k = \frac{q}{4\pi l(T_2 - T_1)} l_n \frac{t_2}{t_1} \qquad (9)$$

where T_1 and T_2 are the wire temperatures at times t_1 and t_2, and q is the heat generated per unit length. The quantities should be taken from the linear part of the time–temperature curve as shown in Fig. 12. The figure clearly indicates the change in slope due to enhancement of conductivity in a nanofluid compared to the base liquid.

The most important observation in their study was a perceptible enhancement in thermal conductivity for vanishing small concentrations. It was reported that at room temperature for toluene–gold nanofluid 3–7% enhancement was observed for only 0.005–0.011% volume fraction, while the same for water–gold nanofluid was observed to be 3.2–5% for vanishing small concentration of 0.0013–0.0026% volume fraction. It was found that the main reason for such enhancement was the small size (∼10–20 nm) of the particles. The enhancement was higher for water-based nanofluids due to use

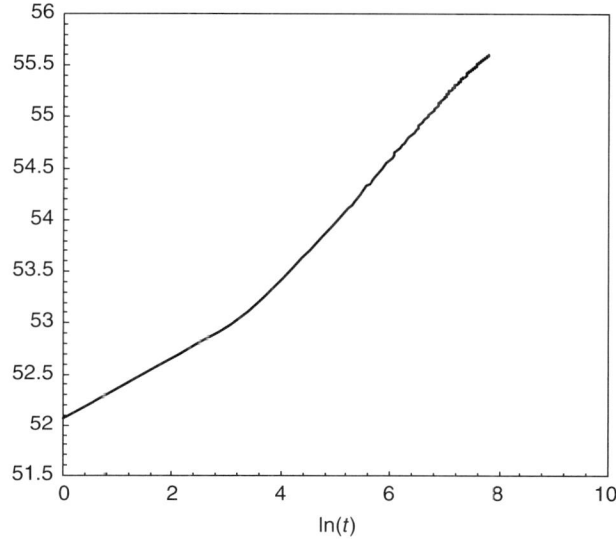

FIG. 12. Temperature variation of platinum wire with time [63].

of bare particles and lower for toluene-based nanofluids where the nanoparticles were protected by a monolayer thiolate coating to prevent agglomeration. Another important observation of their study was the relatively lower conductivity of water–silver nanofluids. It clearly showed that even though silver is higher in conductivity it gave lower enhancement due to its relatively larger size (~60–80 nm) compared to the gold particles (10–20 nm). This indicates that the particle size can override the particle conductivity or concentration effects. One factor has to be kept in mind for hot wire measurement particularly with metallic nanofluids. The wire has to be electrically protected by a thin layer (2–5 μm) of insulation that will prevent electrical interaction with the electrically conducting nanofluids but do not affect the heat transfer [22]. In both the studies (Lee et al. [22] and Patel et al. [38]), the apparatus were tested against standard fluids such as water, ethylene glycol, and toluene for validation of the measurement technique. Liu et al. [64] also confirmed the tremendous potential of metallic nanofluids by showing an enhancement of about 23% at as low concentration as 0.1% volume of copper nanoparticles of 75–100 nm size, suspended in water without any surfactant. They also showed that the thermal conductivity enhancement decreases with time due to sedimentation. Xie et al. [65] made a study on the dependence of effective thermal conductivity of nanoparticles–fluid mixture on the nature of the base fluid. Nano-sized α-Al_2O_3 was dispersed in deionized water, glycerol, ethylene glycol, pump oil, ethylene glycol–water mixture, and glycerol–water mixture, and the study was conducted. It was found that thermal conductivity ratios decrease with an increase in thermal conductivity of base fluid.

Hong et al. [66] achieved enormous magnitude of increase in the thermal conductivity of nanofluids of Fe nanoparticles of 10 nm size suspended in ethylene glycol. They got an enhancement of 18% for just 0.55% volume fraction. They also showed that the sonication of the nanofluid has an important effect on the thermal conductivity of nanofluid, indirectly proving the particle size effect on the thermal conductivity of nanofluids. In another study [67], they showed separately the effect of sonication time on thermal conductivity and cluster size of nanoparticles. They showed that the thermal conductivity of Fe (10 nm) in ethylene glycol nanofluid increases with sonication time, till 50 min of sonication, after which it is found to be remaining constant. Similarly, upto 50 min after sonication, cluster size of nanoparticles was found to be increasing from around 1.17 μm at beginning to around 2.3 μm, after which it was stabilized at this size. They also observed an enhancement of 18% for a volume fraction of 0.55%. Murshed et al. [68] have developed a new technique to measure the thermal diffusivity of liquids directly using transient double hot-wire technique. They measured the thermal conductivities as well as thermal diffusivities of TiO_2, Al_2O_3, and Al nanofluids in ethylene glycol and engine oil and showed that the measured thermal diffusivity of these nanofluids is 5–10% higher than that calculated from the measured thermal conductivity and taking volumetric average of the specific heats. They also showed that the measured thermal diffusivity value for TiO_2 nanoparticles of 10×40 nm size suspended in ethylene glycol was upto 30% higher for 5% volume fraction, whereas that for Al nanofluids (80-nm-sized particles in ethylene glycol) was upto 50% higher for the same volume fraction, indicating high thermal diffusivity enhancements with nanoparticle suspensions. Also, the enhancement at lower concentrations is found to be relatively higher. Chopkar et al. [69] have obtained very promising features of metallic nanofluids. They observed 120% enhancement for just 2% volume fraction of Al70Cu30 particles in ethylene glycol. They also observed a strong nonlinear dependence of thermal conductivity enhancement on particle volume fraction. In addition, they have shown a strong effect of crystalline size on the enhancement.

C. CARBON AND POLYMER NANOTUBE NANOFLUIDS

The largest magnitude of enhancement of effective thermal conductivity in nanofluids was observed in a subsequent study from the ANL group by Choi et al. [37,42]. Figure 13 shows the enhancement of thermal conductivity of

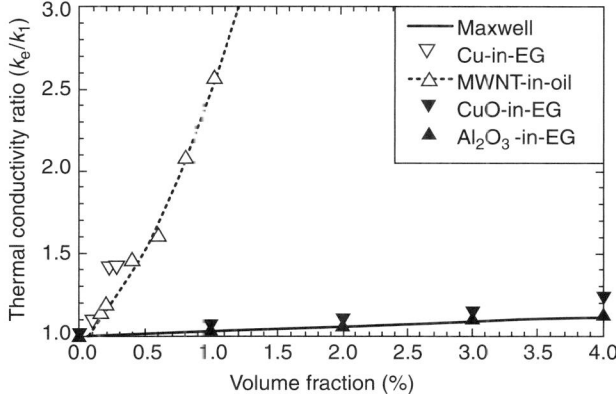

FIG. 13. Enhancement of thermal conductivity of MWNT against volume fraction [42].

MWNT – engine oil nanofluids against nanotube volume fraction (in %) with a phenomenal 159% increase in thermal conductivity with just 1% volume fraction of the nanotubes.

This hike in enhancement is interesting. With polymer nanotubes also, similar enhancement was reported by Biercuk et al. [70]. It is also interesting to note that while the previous measurements with ceramic as well as metallic nanofluids gave an approximately linear relationship, the same with nanotubes is strongly nonlinear. The reason for the abnormal rise of enhancement and the nonlinear behavior is yet to be explained because there are unique features in it with respect to temperature which will be discussed later. However, one can look at two facts: first, the thermal conductivity of CNTs is very high ($\sim 3,000$ W mK^{-1}) and second, the nanotubes have got very high aspect ratio ($\sim 2,000$). We shall indicate the implications of their aspect ratio while requiring the possible theories on thermal conductivity of nanofluids.

Getting a stable CNT nanofluid is tricky. One of the major issues here is the hydrophobically of CNTs. Hence most of the CNT based nanofluids studied were higher treated chemically to reduce hydrophobically or suspended with appropriate surfactant.

Xie et al. [71] have measured the thermal conductivity of multiwalled CNTs of 15 nm average diameter and 30 μm length suspended in water, ethylene glycol, and decene. The suspensions in water and ethylene glycol were without any surfactant but coated with oxygen-containing functional groups, whereas they were suspended in decene with the help of oleylamine as surfactant. It was found that there was more enhancement for same volume fraction in the fluid having lower thermal conductivity. Maximum enhancement in the thermal conductivity was found in decene which was 20% for 1% volume fraction of CNTs. Also it was found to be increasing linearly with volume fraction. However, the enhancement is far small as compared to that of achieved by Choi *et al.* [42]. Assael *et al.* [72] measured the thermal conductivity of MWCNTs as well as double-walled CNTs (DWCNTs). The thermal conductivity of C-MWNT of around 130 nm average diameter and 40 μm average length was found to be 34% for 0.6% volume fraction, whereas that of C-DWNT was found to be 8% for 1% volume suspension in water. This reduction in enhancement compared to Choi *et al.* [42] may be due to the reduced aspect ratio of the CNTs they used. A good review of their work is also presented by Assael *et al.* [73]. Hwang *et al.* [74] have also got similar results for C-MWNT suspensions in water as well as ethylene glycol. Liu *et al.* [75] measured the thermal conductivity of C-MWNTs of 20–50 nm in diameter. They observed an increase of 12.4% in the thermal conductivity of CNT suspension in ethylene glycol for 1% volume whereas 30% enhancement in the CNT suspension in synthetic oil for 2% volume. Recently, Honga *et al.* [76] have shown the impact of magnetism on enhancement of thermal conductivity of CNT nanofluids, where they observed an enhancement of about 50% for just 0.01 wt% of CNTs and 0.02 wt% of magnetically sensitive Fe_2O_3 nanoparticles suspended in water, after applying a magnetic field. It was also seen that the enhancement reduces after some time, possibly due to the agglomeration of the particles under the magnetic field. Zhu *et al.* [77] have measured the thermal conductivity of graphite nanofluids and observed an enhancement of 34% for 2% volume fraction of graphite. This nanofluid is found to be as good as short-length CNT suspensions from a thermal conductivity point of view. The nanofluid was made of 15-nm-thick and 50–100 nm-sized graphite of flake shape suspended in water. Hwang *et al.* [61] have measured the thermal conductivities of MWCNT (10–30 nm diameter and 10–50 μm length), fullerene (10 nm), CuO (33 nm), and SiO_2 (12 nm) suspended in water, ethylene glycol, and oil. They observed enhancements of about 9% for 0.5% volume fraction of MWCNT in water, 5% for 1% volume of CuO in water, and 3% for 1% volume fraction of SiO_2 in water. Enhancements in other fluids were of the similar magnitude, except for that of fullerene which gave an enhancement of 6% for 5% volume fraction in oil, whereas a there is a reduction of 2% for 1% volume fraction in water, as thermal conductivity of fullerene is lower than water but higher than oil. Yang *et al.* [78] measured the thermal conductivity of MWCNT suspended in polyalphaolefin (PAO) and showed a profound effect of aspect ratio of CNTs. The enhancement increases rapidly with the increase in the aspect ratio of CNTs. Also, they have reported very high enhancement of 100% for just

0.25% volume fraction and 200% for just 0.35% volume fraction. Shaikh et al. [79] have measured the thermal conductivity and thermal diffusivity of CNTs, exfoliated graphite (EXG), and heat-treated nanofibers (HTTs) in PAO oil using modern light-flash technique. They observed a maximum enhancement in thermal conductivity of CNT suspensions which was 161% for 1% CNT loading, whereas it was 131% and 103% for EXG and HTT suspensions, respectively, for the same loading. In catranst to the findings of Murshed et al. [68], they observed a less enhancement in thermal diffusivity of these nanofluids as compared to the enhancement in thermal conductivity. They observed thermal diffusivity enhancement of 136, 112, and 88% in the nanofluids of CNT, EXG, and HTT, respectively, for 1% loading.

VIII. Temperature Effect

Nanofluids are unique in many respects such as their sensitivity to change. One important contribution on nanofluid in this direction was from Das et al. [21] who discovered a very strong temperature dependence of nanofluids with the same Al_2O_3 and CuO particles as used by Lee et al. [22]. Before explaining their results, it is interesting to explain the particular conductivity measuring apparatus used by them. This method was developed by Czarnetzki and Roetzel [80] for liquids. The method has the advantage that it can work with extremely precise mean temperature in the cell, requires small volume of liquid, and very low possibility of natural convection. But the most distinctive feature of the method is that it is purely a thermal method with electrical components far away from the fluid. It avoids any magnetic or electrical interaction between the particles and the applied power for measurement. The test cell used for this method is shown in Fig. 14. It is a cylindrical cavity filled with fluid with temperature measuring thermo couples at its end faces and the center. At the ends through the Peltier heating elements in conjunction with a function generator gives temperature oscillation that diffuses through a reference layer and finally propagates through the fluid.

Assuming a one-dimensional heat flow, the equation can be given by:

$$\frac{1}{\alpha}\frac{\partial T}{\partial t} = \frac{\partial^2 T}{\partial x^2} \tag{10}$$

The periodic boundary conditions can be given by

$$T(x=0, z) = T_m + T_o \cos(z + G_o) \tag{11}$$

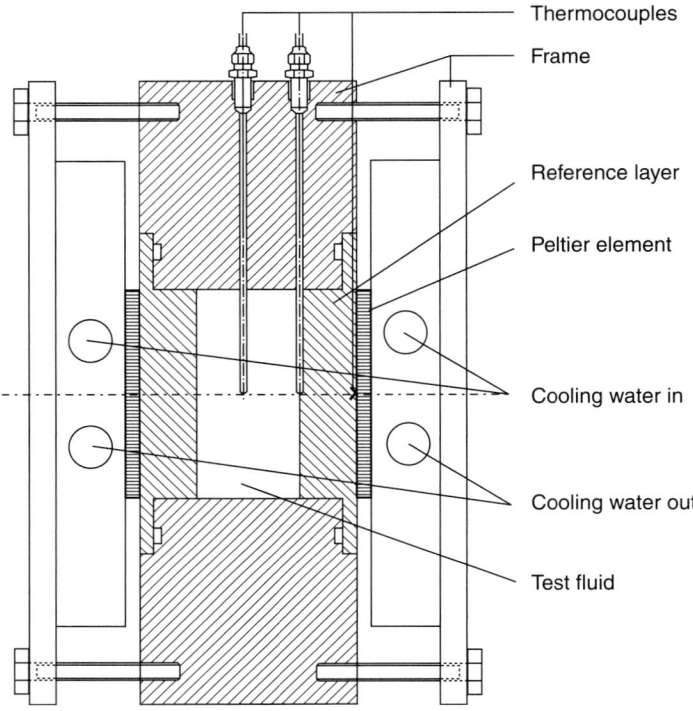

FIG. 14. Test cell for temperature oscillation technique [21].

$$T(x = L, z) = T_\mathrm{m} + T_\mathrm{L}\cos(z + G_\mathrm{L}) \tag{12}$$

where

T_m = fluid mean temperature;
T_o, T_L = amplitude of temperature oscillation at the two end faces;
t = time;
x = axial distance; and
G_o, G_L = phase shift of the input oscillation.

The solution can be found as

$$\text{Amplitude attenuation}: B^* = \frac{2T_\mathrm{L} e^{iG_\mathrm{L}}}{T_\mathrm{L} e^{eG_\mathrm{L}} + T_\mathrm{o} e^{iG_\mathrm{o}}} \cos\left[\frac{1}{2}\left(\frac{iw}{\alpha}\right)^{1/2}\right] \tag{13}$$

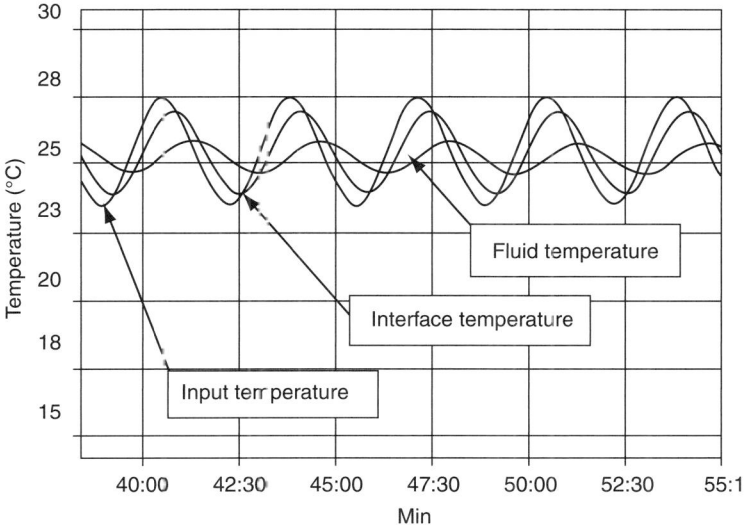

FIG. 15. Temperature oscillation plot [21].

$$\text{Phase shift: } \Delta G = \arc \tan \left[\frac{I_m(B^*)}{R_e(B^*)} \right] \qquad (14)$$

Thus, the derivation shows another advantage of the process which is the possibility of determination of thermal diffusivity (α) either from the amplitude attenuation (B^*) or from the phase shift (ΔG). In fact, what would be more desirable is to determine it from either of them and cross-check it with the other one. A typical temperature oscillation plot is shown in Fig. 15.

It should be mentioned here that temperature measurement in the fluid element can give only thermal diffusivity. To measure thermal conductivity, the measurement across the reference layer has to be made. In this case, the accuracy of the measurement will be limited by how accurately the thermal conductivity of the reference layer (usually stainless steel or copper) is known. The amplitude of temperature oscillation is usually kept to 2–3°C to avoid convection. It is important to see that the method was able to reproduce the results of Lee et al. [22] as well as standard conductivity values of water.

Using the above method, the measurement of the thermal conductivity of oxide nanofluids over a range of temperature 21–51°C was carried out. The results were astonishing. It was found almost three times increase in enhancement (10% became about 30%) of copper oxide and alumina nanofluids as shown in Figs. 16 and 17, respectively.

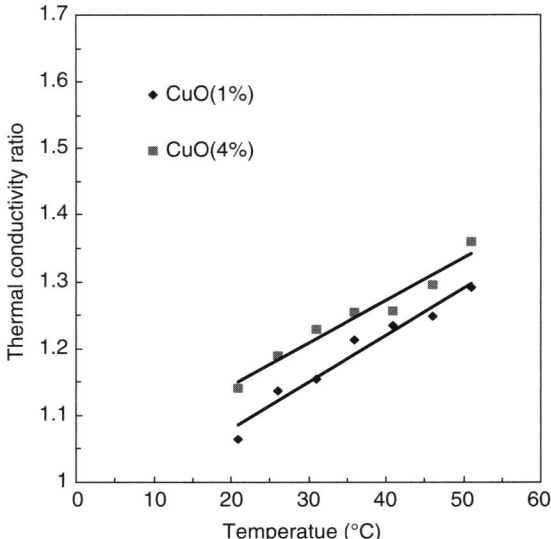

FIG. 16. Enhancement in thermal conductivity of copper oxide–water nanofluid with temperature [21].

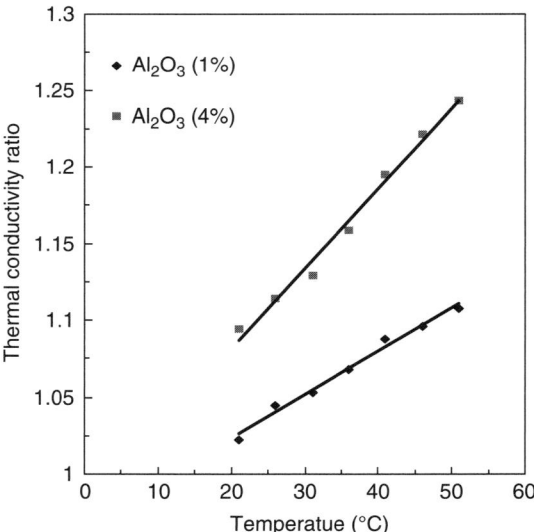

FIG. 17. Enhancement in thermal conductivity of aluminium oxide–water nanofluid with temperature [21].

One important observation here is that in contrary to what was observed at room temperature [22], both Al_2O_3 and CuO do not agree with Hamilton–Crosser [13] model because the model is not sensitive to temperature at all over this small temperature range. Thus, it can be said that for each nanoparticle–fluid combination, there is a threshold temperature below which it behaves like normal suspensions but above this it behaves as a nanofluid. It was concluded that the agreement of Al_2O_3–water nanofluid with Hamilton–Crosser model at room temperature was purely accidental because of its larger particle size These results have revolutionized the concept about nanofluids from an application point of view because much larger thermal conductivity in the heated state makes them "smart fluid" at elevated temperature. This also indicates that some kind of particle movement (probably that of Brownian type) dramatically changes with temperature which must be taking place inside the fluid.

The work by Patel *et al.* [38] reconfirmed this fact. Even for a very dilute nanofluid of completely different particle–fluid combination, a notable temperature effect was observed. Figures 18 and 19 show these effects.

It is again interesting to note that the temperature enhancement of the fluids conductivity goes up from 3% to 9% with concentrations of the order

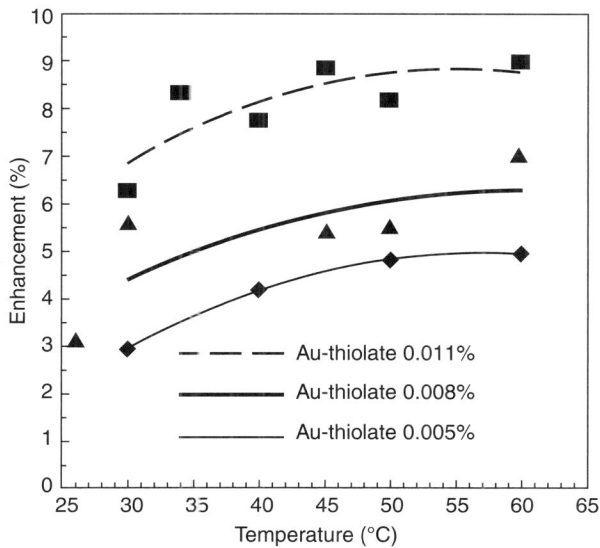

FIG. 18. Percentage enhancement in thermal conductivity of gold–toluene nanofluid with temperature [38].

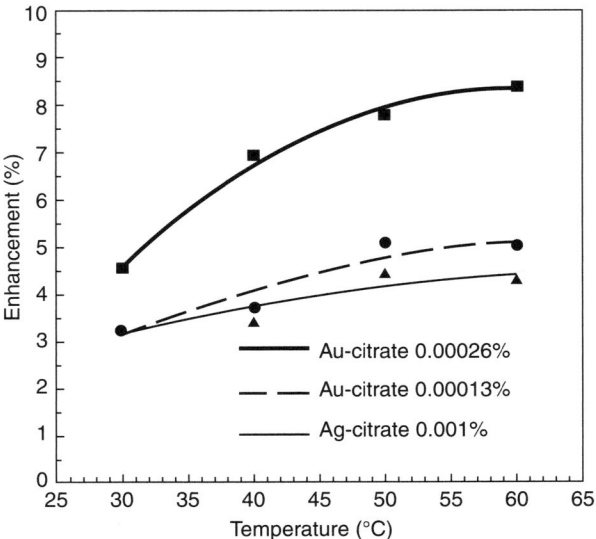

FIG. 19. Percentage enhancement in thermal conductivity of gold–water nanofluid with temperature [38].

of ~0.01% (coated particles in toluene) and ~0.0002% (bare particles in water). Temperatures effects were in general nonlinear in nature; however, over a small range, it can show linear behavior. The other important feature of the study was that over the entire range of temperature, silver which has got higher conductivity gave lower conductivity enhancement due to its large size (~60–80 nm) compared to gold (~3–20 nm) particles used. The slope of the temperature–conductivity curve is also the lowest for water–silver nanofluids, indicating a lower temperature effect closer to normal suspension behaviors.

Chon et al. [81] also confirmed the temperature effect got by Das et al. [21] with similar nanofluid. They also showed the inverse dependence of thermal conductivity enhancement on the particle size. These findings were also confirmed by Li et al. [82], who showed a little stronger effect of temperature for the same nanofluids as that used by Das et al. [21]. They observed an increase in the enhancement from 8% to 20% for a temperature rise from 28°C to 36°C for 2% volume fraction of alumina nanoparticles suspended in water and 30–40% for 2% volume fraction of CuO nanoparticles suspended in water in the same temperature range. Murshed et al. [83] have also observed the temperature effect for aluminium–engine oil nanofluids (80 nm particle size). They found an enhancement of 20–36% for a temperature rise of 20–60°C for 3% volume fraction. Recently, Li and Peterson [84] have shown from the experimental investigation of alumina (36 nm and 47 nm) and water nanofluids that the effects of different parameters such as particle size, suspension temperature, particle volume fraction, and particle material–fluid combination on the thermal conductivity of nanofluids are different for different combinations and there lies a need to optimize it for getting best performance out of a particular combination of solid and liquid phases. In another novel study, Yang and Han [85] prepared a nanoemulsion of water in perfluorohexane (FC72) with average size of water droplets to be 9.8 nm. They found the thermal conductivity of the nanoemulsion to be 52% higher than that of FC72, whereas the effective-medium theory prediction was around 10–15%.

However, recently, Zhang et al. [86] observed no enhancement in the thermal conductivity as well as thermal diffusivity of nanofluids above that of predictions of the Hamilton–Crosser model, neither did they observed any temperature effect as observed by Das et al. [21]. They speculated that the enhancement observed in literature might be due to the possible leakage of electricity in the measurement cell (while using transient hot-wire (THW) instrument for thermal conductivity measurement), thereby inducing error in the thermal conductivity measurement. Zhang et al. [86] have also reached the same conclusions from the measurement of thermal conductivity and thermal diffusivity of Al_2O_3, ZrO_2, TiO_2, and CuO nanofluids. Putnam et al. [87] have also measured the thermal conductivity of C60–C70 fullerenes in toluene and suspensions of alkanethiolate-protected Au nanoparticles to maximum volume fractions of 0.6% and 0.35% using optical beam deflection technique and observed no significant enhancement in thermal conductivity of these nanofluids above that predicted by effective-medium theory. Venerus et al. [88] have also reported no enhancement in thermal conductivity as well as any temperature effect for gold–water nanofluids as well as alumina–oil nanofluids, where they used forced Rayleigh scattering technique to measure thermal conductivity of nanofluids. Yang et al. [85] have also observed no enhancement as well as no effect of temperature for Bi_2Te_3 nanorods of 20 nm diameter and 170 nm length suspended in oil as well as FC72, as compared to the predictions of effective-medium theory for thermal conductivity.

These studies bring out some scepticism in the measurement method used in literature as well as the basic phenomenon itself. However, an overwhelming majority of investigation including the studies in convection and boiling (discussed later) indicates that the nano-effect in nanofluids is not measurement inconsistency alone.

IX. Theories on Thermal Conductivity of Nanofluids

Right from the time when Choi [16] invented nanofluids, there has been a continuous effort to look for the causes of the "anomalous" increase in thermal conductivity of nanofluids. Starting from simple Brownian motion to complicated fractals, by revisiting Maxwell and assuming formation of liquid layer around particle, many propositions have been tried. During the past 3–4 years, this effort of theorizing nanofluid behavior has intensified. However, it appears that the truth is still at least half revealed.

The traditional theories of Maxwell [12], Hamilton and Crosser [13], Wasp [14], Bruggeman [89], and Bonnecaze [90,91] explained the thermal conductivity enhancement of usual slurries and suspensions quite extensively. The basic model of Maxwell [12] was extended by the following investigators by including the effect of shape [13], particle interactions [56–58,92,93], and particle distribution [94]. The Bruggeman [89] model has got the advantage of being valid for a wide range of concentration.

$$k_{\text{eff}} = (3\phi - 1) + [3(1 - \phi) - 1]k_{\text{f}} + \sqrt{\Delta} \tag{15}$$

$$\sqrt{\Delta} = (3\phi - 1)^2 k_{\text{p}}^2 + [3(1 - \phi) - 1]^2 k_{\text{f}}^2 + 2[(2 + 9\phi)]k_{\text{p}}k_{\text{f}} \tag{16}$$

Generally, Maxwell's method works well for low thermal conductivity ratio between the solid and the fluid ($\sim$10). Also, all these effective-medium theories consider the particle as point mass sources without accounting for their sizes. As far as the enhancement of the conductivity with concentration is concerned, nanofluid shows qualitatively similar trends as predicted by the classical models. This is evident from the near linear variation of conductivity enhancement reported by Lee *et al.* [22], Das *et al.* [21], and Patel *et al.* [38]. However, enhancement is much higher compared to what is predicted by these models. Yu *et al.* [93] shows that the model predicts less than 5% enhancement for 1% volume fraction of particles while that can be as high as 40–50% in nanofluids. Figure 20 shows the thermal conductivity enhancement of nanofluids for different particle to fluid conductivity ratio predicted by the classical models as given by Yu *et al.* [93]. Figure 20 shows the difference between the measured values of conductivities and that predicted by the Maxwell model.

As has been already pointed out, the temperature effect makes this deviation more even for the oxide nanoparticles that appears to be close to the Maxwell model in this figure.

Lee *et al.* [22] and Wang *et al.* [30] have also shown that even at room temperature, the results for oxide nanoparticles start deviating from the Maxwell's theory at lower concentration. This gives an impression that

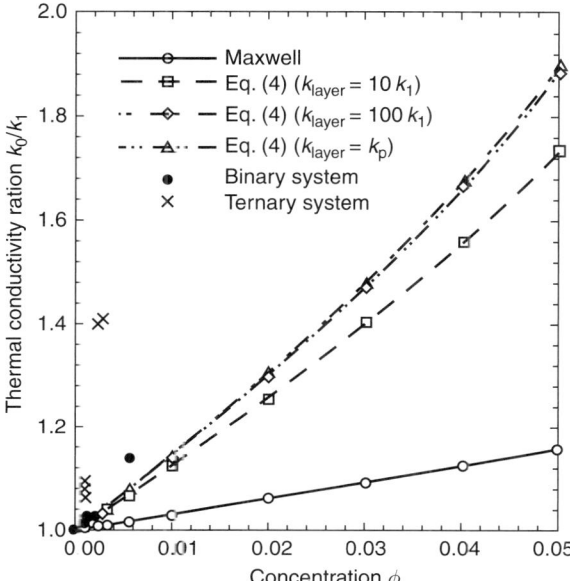

FIG. 20. Thermal conductivity enhancement as shown by Yu and Choi [93].

particle movement (and hence the mean free path) is important for effective thermal conductivity enhancement.

The failure of the classical theories to predict nanofluid behavior gave rise to hypotheses regarding the mechanism of heat transfer in nanofluids. Wang et al. [30] attributed the enhancement due to particle motion, surface action, and electrokinetic effects. Electrokinetic forces may be Van der Waals forces, electrostatic forces, or micro-stochastic forces. They indicated that electric double layer [80] can play a significant role in it. The hydrodynamic force in the form of microconvection can also be a cause of the enhancement. Their TEM studies showed some chain structures and they suggested that these structures can also be the cause of conductivity enhancement.

A serious look at the various possibilities of enhancement mechanism was focused by Keblinski et al. [95]. The mechanisms considered by them were:

1. Brownian motion
2. Ballistic conduction
3. Liquid layering
4. Particle clustering

At the very first sight even though the Brownian motion appears to be a probable mechanism, it was rejected to be the major responsible mechanism by them from a timescale study. It must also be mentioned here that the studies of Wang *et al.* [30] also showed that Brownian motion does not contribute significantly. Keblinski *et al.* [95] showed that liquid layering around the particle "could give" the path for rapid conduction. This has given rise to a number of studies on this theory later, but the existence and the real nature of such layers is still an open question The other mechanism suggested by them was particle clustering and the mechanism of ballistic heat transport gains significance due to the fact that the phonon mean free path is of the order of nanoparticle dimensions. The molecular dynamics simulation (MDS) carried out by them shows the ballistic nature of energy transport inside the nanoparticles. They have shown that the nature of the autocorrelation function is monotonic decay inside the liquid and oscillatory inside the solid with a period that corresponds to the residence time of the phonon inside the particle. However, they concluded that even ballistic transport theory does not support the type of conductivity enhancements observed in the nanofluids. The theory of particle clustering seems to be promising at the first sight, but it also suffers from a number of deficiencies. The TEM studies do not show any loosely packed cluster which Keblinski *et al.* [95] investigated. The clustering should increase the effective particle size and thus help in sedimentation that was not observed in practice. Also, large temperature effect cannot be explained in terms of clusters. The clustering effect can be shown in Fig. 21.

It shows that while a separate loosely packed cluster can give thermal conductivity enhancement as high as 10 times that of the base fluid, this reduces exponentially with loosely packed particles in contact such as simple

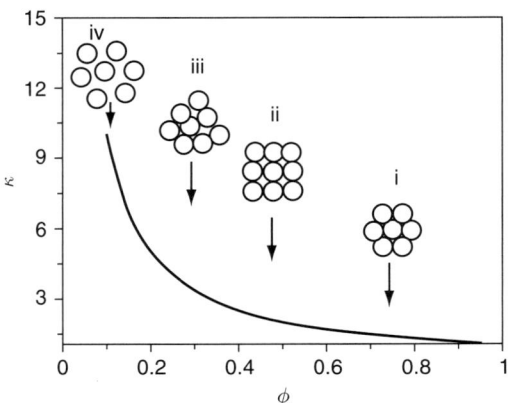

FIG. 21. Particle clustering effect on heat transfer augmentation. (i) closely packed fcc arrangement of particles, (ii) simple cubic arrangement, (iii) loosely packed irregular structure of particle in physical contact, and (iv) clusters of particles separated by liquid layers thin enough to allow for rapid heat flow among particles.

cubic packing or closely packed face-centered cubic (FCC) structure. Liquid layering theory was shown to be promising, but it also uses the adjustable parameter of thickness of liquid layer that is yet to be supported by experimental investigation. The transport at nanoscale is obviously to be modeled with the relevant theories and the nanoscale modeling with Boltzmann transport equation (BTE) appears to be appropriate. However, the solution of BTE with nanoparticles in a host medium by Chen [96] indicates a lowering of effective conductivity for nonlocal, nonequilibrium conduction rather than enhancement and hence such microscopic treatment also fails to predict the observed enhancement in nanofluid conduction.

Wang et al. [97] approached the theory from the route of fractal geometry. They followed an approach Pitchumani et al. [98] took for fibrous composites. They included surface adsorption as well as particle conductivity approach in contrast to the bulk conductivity used by other models. Use of particle conductivity seems to be logical due to the fact that all properties including thermal conductivity suffer from size effect at nanoscale of dimensions. The results indicated that for water–CuO nanofluids, the fractal model shows quite good promise when adsorbing is included in the analysis while it under predicts the enhancement in absence of adsorption. It must be said that their comparison was limited only to the CuO–water nanofluid data measured by them with a quasi steady-state method. They neither compared with the data for metallic nanofluids nor did show any effect of temperature. The fractal model is actually an extension of the Maxwell–Garnett model [55] with the inclusion of fractal dimension. This model also depends on some kind of clustering of particle which again goes against a large number of observations in which majority of the particles were found to be in perfectly dispersed state [21,22,38]. Ma et al. [99] have shown the high potential of nanoparticle suspension in liquid metal through their model, where they have incorporated the modified thermal conductivity of nanoparticles in Maxwell model [55].

A novel approach in the modeling of nanofluid was by Xue [100]. He considered the particles to be ellipsoids interacting with spherical fluid particles. The model utilizes a field factor approach with depolarization factor and effective dielectric constant. The main assumption of this model is the existence of a conducting shell that is equivalent to liquid layering theory. The surface activity is taken to be the responsible factor and he showed that for specified values of the shell thickness (~ 3 nm) and shell conductivity ($k_1 = 5 \, \text{W m}^{-1} \text{K}^{-1}$ for CNT–oil nanofluid and $k = 2.1 \, \text{W m}^{-1} \text{K}^{-1}$ for Al_2O_3–water nanofluid), the theory matches with the measured values. However, it must be said that it uses these two adjustable parameters to match with the experimental values and no evidence of such layer and the values of either the thickness or the conductivity of such shell has yet been found; hence, the model requires approximation for these two fitting parameters, the nature of variation of which is unknown. Thus the model is insufficient to predict the enhancement of any unknown combination of particles and fluid. However, the model is a remarkable achievement in the sense that for the first time one model was presented which can predict both the low enhancement of conductivity in Al_2O_3-based nanofluids and a very high enhancement in CNT-based nanofluids. Being able to club both the extreme cases into one, theoretical model should not be undermined and the model can be of great importance for development of thermal conductivity models in the future. Another model in the same line as above but without considering the liquid layering was proposed by the same author [101] for predicting the thermal conductivities of CNTs. The model is found to be working fairly well for predicting the thermal conductivities of CNT suspensions. Similarly, one more model [102] considering only temperature function and liquid layering was given by the same author for predicting the thermal conductivities of nanoparticle suspensions. The model is found to predicting the thermal conductivities of CuO particle suspensions in water as well as ethylene glycol; however, both of the above models carry the same drawbacks as that of the model given in [100].

The interfacial layer model was also used by Yu and Choi [93]. They decided to use the classical Maxwell model as the base model in which the liquid layer around the particle that has a "solid-like" behavior is introduced. This gives an increase in the effective concentration. Here also, they used a liquid layer thickness and a layer conductivity that have been adjusted for matching. However, they used equivalent particle conductivity as given by Schwartz [103] for calculation. The range of the liquid layer thickness which could predict the results was ~2 nm. They claimed that the model was not very sensitive to the conductivity of the layer and a value of layer conductivity which is 10 times the fluid conductivity gives a good match. The results for copper oxide with ethylene glycol and copper particles (3 nm) in ethylene glycol were quite close to experimental values, but for copper particles with surfactant where maximum enhancement was observed (40% with 0.3% volume fraction), the model failed miserably. Like Xue [100], this model also uses adjustable parameters and predicts well under certain conditions. Even though the model does not really contribute new to the understanding of the heat transfer mechanism, it reiterates the possibility of liquid layering and lack of sensitivity to layer conductivity that can be useful for modeling. Xie *et al.* [104] modeled the thermal conductivity of the liquid layer and incorporated it in effective thermal conductivity of nanofluid. But they too have validated it against only a few experimental results, that too considering a fixed nanolayer thickness of 2 nm; while the nanolayer thickness may be expected to be different for different combinations of liquid and solid, the physics of the formation of such layers is also under all the above studies. The major drawback of the works where effective thermal conductivity of nanofluids is tried to explain through only liquid layering is that the size of layer is assumed to be very high, which does not have any experimental validation and the thermal conductivity of liquid layer is taken to be as high as the solid's thermal conductivity. The only experimental proof of liquid layer shows it to be only a few (three) atomic diameters thick [105]. Also in a numerical work, Xue et al. [106] have confirmed this fact from a fundamental point of view. Using molecular dynamics simulation, they have shown that the effect of high surface energies on nanoparticles and the interactions between the solid and liquid molecules cannot affect the properties of the surrounding liquid for longer than five atomic distances. The crystal-like atomic configuration and hence the enhanced mechanical and thermal properties were found to be getting faded within five atomic distances, which is much smaller than the layer thickness assumed in the above studies. Leong *et cl.* [107] gave a model in the similar line; however, the liquid layer thermal conductivity is not incorporated in the effective thermal conductivity of particles, but treated separately. It has been validated against many of the experimental results for thermal conductivity of oxide nanofluids with adjustment of variable parameters. The major drawback of the work that tries to explain effective thermal conductivity of nanofluids only through liquid layering is that the size of layer is assumed to be very high, an assumption that has not been experimentally validated, and that the thermal conductivity of the liquid layer is taken to be as high as the thermal conductivity of the solid. The only experimental proof of a liquid layer shows that it is only a few (three) atomic diameters thick [71]. Also in a numerical work, Xue *et al.* [106] have confirmed this finding from a fundamental point of view. Using molecular dynamics simulation, they have shown that the effect of high surface energies on nanoparticles and the interactions between the solid and liquid molecules cannot affect the properties of the surrounding liquid for more than five atomic distances.

Taking effective-medium theory approach, Prasher *et al.* [108] have shown the importance of the fractal clusters in the increased thermal conductivity of nanofluids. Feng *et al.* [109] have also shown the effect of aggregation on thermal conductivity of nanofluids by coupling it with liquid layering and effective-medium theory. By approximating the actual particle volume fraction to effective volume fraction, they have matched their model predictions with some of the experimental data to validate the volume fraction effect.

Another approach has gained momentum in explaining the thermal conductivity enhancement and that is the incorporation of particle motion mainly due to the need to explain temperature effect. Even though it has been started earlier that Brownian motion alone cannot account for this enhancement, a new look at the Brownian motion [110,111] has been given. Based on a drift velocity model, Yu *et al.* [93] have first shown that the collision of particles and the drift velocity can account for a very small part of the enhancement. With a specific example of copper particles in ethylene glycol, they showed that at least the order of the enhancement could be guessed if a nanoconvection in the space between the particles is assumed. This is quite logical because in usual Brownian motion in gases between the particles (molecules), there is only void, while in nanofluids these fluids will be participating in a nanoconvection which may even be set by electrical dipole. They [93] even pointed out the possibility of a Soret effect [112]. This work does not present a complete accurate model but throws light at a very possible mechanism and more importantly tries to model the phenomenon from the fundamental physics without any adjustable parameters. It will be interesting to see the efforts of modeling in this direction.

Xuan *et al.* [113] have presented another model which primarily rests on the paradigm of Brownian motion. They combined the concept of fractals and Brownian motion. The fractal conception was used to incorporate the diffusion-limited aggregation of particles. The whole process has been assumed to have two additive parts: the usual static theory of suspension and the Brownian motion-dominated dynamic part. Naturally the Brownian motion part is temperature-dependent and was proposed to be proportional to $\sqrt{T}$. They did not provide any evidence of the proposed temperature effect against experimental data and the composite conductivity enhancement was proposed as

$$\frac{k_{\text{effective}}}{k_f} = \frac{k_p + 2k_f - 2\phi(k_f - k_p)}{k_p + 2k_f + \phi(k_f - k_p)} + \frac{\rho_p \phi C_p}{2k_f}\sqrt{\frac{k_B T}{3\pi r_c \mu}} \qquad (17)$$

where

μ = viscosity;
T = temperature; and
r_c = radius of the cluster.

While this model for the first time incorporated temperature effect into the model, it suffers from the same serious drawback. The most important shortcoming was that even though the enhancement was proposed as a function of temperature, only room temperature measurement was used for validation. This does not validate the $\sqrt{T}$ variation proposed by them. Moreover, the results of temperature effect [21] clearly show that the model does a poor job in predicting this effect.

The true nature of enhancement as well as temperature effect was recently modeled by two groups using essentially the Brownian motion concept. However, the approach and explanations were different in the models. Jang and Choi [114] have modeled nanofluids based on conduction, Kapitza resistance at particle surface, and convection. In deriving their model, they considered four modes of energy transport: (1) collision between base fluid molecules, that is, thermal conductance of fluid; (2) thermal diffusion in nanoparticles; (3) collision between nanoparticles due to Brownian motion, which was neglected by order of magnitude analysis; and (4) thermal interaction of dynamic nanoparticles with base fluid. Brownian motion produced convection-like effects at nanoscale.

The model including this micro-convection model can be given by Eq. (18). As particle size is decreased, the random motion becomes larger and convection-like effects become dominant.

The effective thermal conductivity was given as

$$k_{\text{eff}} = k_{\text{BF}}(1-f) + k_{\text{nano}} f + 3C_1 \frac{d_{\text{BF}}}{d_{\text{nano}}} k_{\text{BF}} \text{Re}_{\text{dnano}}^2 \text{Pr} f \qquad (18)$$

where

k_{BF} = thermal conductivity of base fluid;
f = volume fraction of nanoparticles;
k_{nano} = thermal conductivity of nanoparticles;
C_1 = proportional constant;
d_{BF} = diameter of base fluid molecules;
d_{nano} = diameter of nanoparticle;
Re_{dnano} = particle Reynolds number; and
Pr = Prandtl number of nanofluid.

This model was able to predict particle size and temperature-dependent conductivity accurately. In the similar lines of the above model, Xu et al. [115] gave a model, which added the convection effects coupled with fractal analysis to the effective conductivity of nanofluid, as derived from the classical model of Hamilton and Crosser [13]. The model was validated against thermal conductivity of some of the oxide nanofluids using an empirical constant.

In an effort to include all observed effects, Hemanth et al. [116] have given a model, which accounts for the dependence of thermal conductivity on particle size, concentration, and temperature. The proposed model has two aspects. The stationary particle model accounts for the geometrical effect of increase in surface area per unit volume with decreasing particle size. It assumes two parallel paths of heat flow through the suspension: one through the liquid particles and the other through the nanoparticles.

Here, direct dependence of thermal conductivity enhancement on volume fraction and inverse dependence of thermal conductivity enhancement on particle diameter have been suggested, which is as follows:

$$k_{\text{eff}} = k_m \left[1 + \frac{k_p \phi r_m}{k_m (1 - \phi) r_p} \right] \quad (19)$$

where

k_m = thermal conductivity of base fluid medium;
k_p = thermal conductivity of particles phase;
ϕ = volume fraction of nanoparticles;
r_m = liquid particle radius; and
r_p = nanoparticle radius.

The moving particle model developed from the Stokes–Einstein formula [46] explains the temperature effect on thermal conductivity enhancement.

$$\frac{k_{\text{eff}}}{k_m} - 1 = c \cdot \bar{u}_p \frac{\phi r_m}{k_m (1 - \phi) r_p} \quad (20)$$

where

c = proportionality constant; and
$\bar{u}_p$ = Brownian motion velocity of nanoparticles.

Here, the effective particle conductivity is taken as $k_p = c.\bar{u}_p$, where $\bar{u}_p$ is the average particle velocity and c is a constant. This model originates from the observation that temperature effect is proportional to Brownian velocity.

The Brownian velocity was calculated from the Stokes–Einstein equation given by Hiemenz and Rajagopalan [46].

$$\bar{u}_p = \frac{2k_b T}{(\pi \mu d^2)} \quad (21)$$

Predictions from the combined model agree with the experimentally observed values of conductivity enhancement of nanofluids with vanishingly small particle concentration.

They also showed by order of magnitude analysis that the value of the constant c is consistent with that predicted by kinetic theory. However, in the above two models, the constant used is empirical one and varies over several orders of magnitude for different combinations of the particle–fluid mixture. Similar approach is adopted by Ren et al. [117] in which they have considered the kinetic theory-based microconvection and liquid layering in addition to the liquid and particle conduction. They too have treated the particles as a gas-like phase and rendered them a velocity derived from the kinetic theory of gases for a monatomic gas molecule. The model is found to be working well for ceramic particles suspension; however, they too have considered a fixed nanolayer thickness of 2 nm. They have also modeled the thermal conductivity of the nanolayer as volume-averaged thermal conductivity of base liquid and particles. Another model in the same vein was presented by Prasher et al. [118]. They modeled the thermal conductivity of solid particles from kinetic theory of gases and incorporated the Brownian motion-based convective contribution to the total heat transport in the effective-medium approach-based thermal conductivity equation. The Brownian motion velocity considered by them in modeling the microconvection around nanoparticles is based on equipartition theorem, whereas the velocity of particles considered for modeling the thermal conductivity of nanoparticles is phonon velocity. For the first time, they have considered the effect of multiparticle convection, that is, the effect of mixing of convective streams arising form the various motion of particles. The model is found to be working well for ceramic particle-based nanofluids for a particular value of constants used in modeling. However, they considered the effect of particle surface resistance for which, only a few experimental data are available. In the next step of the model, they have refined the model which takes into account the interfacial resistance between solid particles and liquid, convection effects due to the Brownian motion of particles, and the effects of thermal and hydrodynamic interactions due to multiparticle interactions. With the constants varying in a narrow range, they have been able to match the model predictions with many of the experimental results from literature. They indicated in their work that the agitation created in the liquid due to the particle movement contributes maximum in the total heat transfer, instead of the direct contribution of moving particles themselves. Prakash et al. [119] added the liquid layer effect to the above model given by Prasher et al. [118] and showed that their model addresses the major issues about thermal conductivity of nanofluids; however, they have not validated it against experimental data from literature.

Recently, it has been tried to model the thermal conductivity of nanofluid empirically. Patel et al. [120] used a new, semi-empirical approach to model the thermal conductivity of nanofluids. The high enhancements are attributed to the increase in the specific surface area as well as Brownian motion-based microconvection. The Brownian motion-based microconvection is modeled with an empiricism in Nusselt number definition. With that, the model is found to be working excellently over a wide range of combination of nanofluids and over a wide range of parameters. Xuan et al. [121] have also given a model in similar vein using superposition principle and the Green–Kubo theorem. They have analyzed the stochastic motion of nanoparticles to obtain the stochastic temperature variation in the nanoparticles. The semi-empirical model, which accounts for the Brownian motion induced microconvection around the nanoparticles as well as for interfacial resistance, matches well with some of the experimental data from literature. Recently, Patel et al. [122] have proposed a cell model, which is also based on the Brownian motion-driven microconvection around nanoparticles, considering the liquid conduction in parallel with the series combination of conduction and microconvection. The model is semi-empirical in nature and found to be addressing all the major issues like nonlinear dependence of thermal conductivity on volume fraction at low concentrations, particle size effect, and temperature effect, and is validated against lot of experimental data from literature.

In the similar way, a completely empirical model is given by Chon et al. [123] in which they have given an empirical correlation for alumina nanofluids by fitting a curve through regression analysis to the existing experimental data. In this modeling, the microconvection around particles is modeled, considering diffusive velocities of the particles with mean free path of the liquid molecules as characteristic length. This is a new approach in modeling the thermal conductivities of nanofluids.

A more logical formulation in modeling has recently been proposed by Patel et al. [124] in which the microconvection and particle conduction are put in series which is parallel with fluid. This is the first model which could predict nonlinear behavior at low concentration as observed by Hong et al. [66] and Zhou and Wang [125] and temperature effect simultaneously.

Bhattacharya et al. [126] also made Brownian dynamics simulation to determine effective conductivity of nanofluids. The simulation results were with in 3% of experimental data for Al_2O_3-EG (ethylene glycol) and nearly in full agreement with Cu-EG. Recently, Xuan and Yao [127] developed a lattice Boltzmann model to investigate the nanoparticle distributions and flow pattern of the nanofluid. The distribution of the suspended nanoparticles was determined by a series of the acting forces and potentials. A set of formulas of expressing these forces was introduced. Among them, the Brownian force was a dominant factor of affecting random displacement and aggregation of the nanoparticles. It was found that the main flow and rising temperature of the fluid can improve the nanoparticles distribution, which is beneficial to energy transport enhancement of the nanofluid. Eapen et al. [128] have carried out a molecular dynamic simulation of dilute suspension of platinum clusters in xenon using linear response theory and showed the thermal conductivity enhancement of nanofluids to be primarily a surface interaction phenomenon. Li *et al.* [129] investigated the role of fluid mixing due to the Brownian motion-induced microconvection, with CFX software using finite volume analysis. They concluded that the nanoparticles, even without taking part actively in the thermal conduction process, can enhance the total heat transfer through micromixing of the fluid. However, Evans *et al.* [130] have shown in an analysis using molecular dynamics simulation that the high conductivity of nanofluids is characterized by clustering effect and the Brownian motion does not have any significant effect on it.

Nan *et al.* [131] presented a simple formula for the thermal conductivity enhancement in CNT composites, which is derived from Maxwell–Garnett model [55] by effective-medium approach. But the model overpredicts the enhancement in the thermal conductivity of CNT suspensions when calculated with typical values of thermal conductivity of CNTs. The same authors have also given a new model [132] by incorporating the interface thermal resistance with an effective-medium approach. However, the model needs data for interfacial thermal resistance values at the surface of CNTs, which is difficult to get for different types of CNTs and their combinations with different solvents particularly due to their hydrophobic nature. Xue [133] has also presented a model with similar approach using an average polarization theory and shown that the suspensions of longer CNTs can give more enhancement, whereas the diameter of CNTs do not have any effect on it. With similar approach, Gao *et al.* [134] have shown a nonlinear dependence on the thermal conductivity of CNT nanofluids, where their model predicts rapid increase in the thermal conductivity enhancement with increase in volume fraction. In the next step of this model, they introduced particle–liquid surface resistance as a fitted parameter and matched the model results with MWCNT as well as single-walled CNT (SWCNT) nanofluids. In yet another model with similar approach, they have compared their model against the data for thermal conductivity of CNT nanofluids as well as oxide nanoparticle nanofluids. Ju and Li [135] have given a model for thermal conductivity of CNT suspensions, where they estimate interfacial thermal resistance and its relationship with functionalization of surface of tubes, which is incorporated in the Maxwell–Garnett model [55]. Strauss *et al.* [136] presented a simple model for nanofluids of CNTs on a periodic lattice with orientation averaging and no adjustable parameters. The model is able to predict the experimentally observed thermal conductivities of CNT nanofluids with a series and parallel combination of thermal resistances.

Thus, it can be said that a variety of models and mechanisms, depending on extension of classical theory, liquid layering, particle aggregation, and particle movement, has been tried to explain the nanofluid behavior. However, the success of these models has been very limited and no clear picture has emerged till recent times. Only the very recent breakthrough from the present authors [114,116] has started giving more realistic picture explaining all major effects. Quite often, these models require adjustable fitting parameters which are difficult to justify. However, these works have definitely shed light on the various mechanisms of thermal transport in nanofluids. With CNT, the story is quite different and heat percolation through network of CNTs appears to be a major contributor.

X. Convection in Nanofluids

A. Forced Convection In Nanofluids

As seen in the review above, several studies are available in the area of thermal conductivity of nanofluids, but it must be understood that the thermal conductivity enhancement in nanofluid is only a "necessary" condition for its usefulness. The "sufficient" condition comes from the convective studies of these fluids.

Before implementing these fluids as coolants in applications such as engine cooling, electronic chip cooling, and heat exchangers, a thorough, systematic scrutiny is necessary in the convective heat transfer properties of them. However, viscosity of the coolant liquid is one among the dominant properties as it governs the pressure drop and the consequent pumping power involved during flow applications and thus needs a prior investigation. The prime question with respect to the viscosity of nanofluids is whether nanofluids are Newtonian fluids or are shear rate-dependent. Pak and Cho [137] were the first to address this issue. They were entitling the fluid not as nanofluids but as "dispersed fluid with submicron particles." Using the standard Brooks field viscometer, they found that for γ-Al_2O_3 and TiO_2 particles of 13 nm and 27 nm average size suspended in water, the suspensions are Newtonian at very low particle volume fractions and start showing shear thinning behavior (i.e, decrease of viscosity with shear rate) with an increase in particle volume fraction. However, a difference in the results was observed with different nanofluids. While water–Al_2O_3 nanofluid started showing shear thinning behavior at 3% particle volume fraction, the water–TiO_2 nanofluid showed it at 10% particle volume fraction onward. The experimental result was compared with that of a theoretical model. The model by Batchelor [138] is applicable for dilute fluids dispersed with solid sphere, which is given by

$$\mu_r = 1 + 2.5\phi_p + 6.2\phi_p^2 \qquad (22)$$

where μ_r is the relative viscosity of the suspension and ϕ_p is the particle volume fraction. Often a simplified form of this equation known as

Einstein equation may used to estimate the viscosity of dilute suspensions, which is given as

$$\mu = \mu_L(1 + 2.5\phi_p) \tag{23}$$

where μ_L is the liquid viscosity.

Substantial increase in the viscosity of the two nanofluids was observed and the Batchelor model seemed to fail completely for these fluids although the volume fractions of the particles were in the range of the applicability.

They also found that with the increase in temperature, the viscosity of nanofluids decrease, following the same trend as base liquid but the value of viscosity is much higher.

Lee et al. [139] measured the viscosity of water and ethylene glycol-based nanofluids with Al_2O_3 nanoparticles and made similar observation that the relative viscosity of nanofluids increases substantially which can offset the advantage of heat transfer. Das et al. [139] measured the viscosity of water–Al_2O_3 nanofluid and showed a shear rate-independent viscosity. This follows the Newtonian theory for particle concentration up to 4%. Similar behaviors were observed at different temperatures also (Fig. 22).

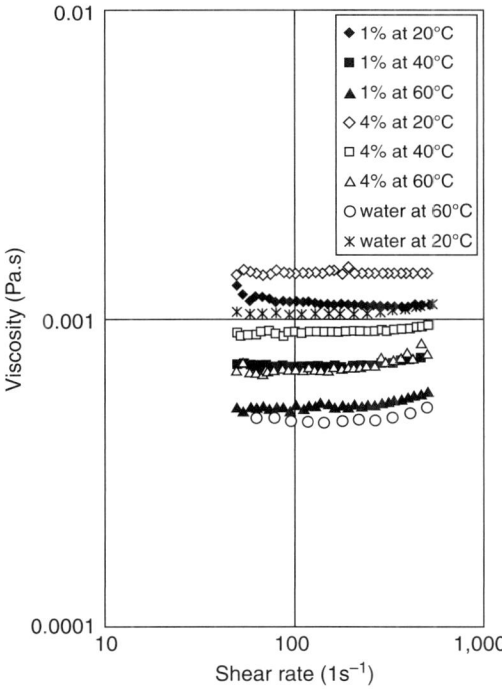

FIG. 22. Dynamic viscosity of nanofluid and pure water at different temperatures [139].

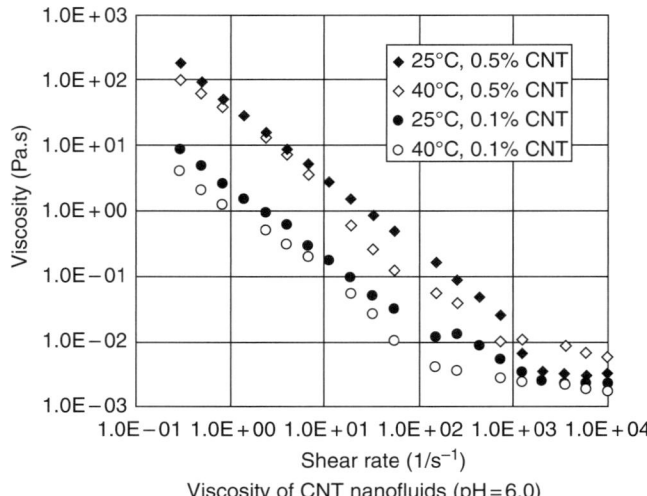

FIG. 23. Viscosity of CNT-containing nanofluid Ding et al. [140].

The CNT-containing nanofluids are quite different in behavior not only with respect to thermal conductivity but also with respect to viscosity. The study by Ding et al. [140] on aqueous CNT-containing nanofluids showed interesting linear shear thinning behavior of the nanofluid at lower shear rate. Since they used gum arabic as stabilizing agent, they also measured the viscosity of the base water with gum arabic which showed nonlinear behavior different from the CNT–water nanofluid. Figure 23 shows the viscosity of CNT-containing nanofluids.

It must be mentioned here that the traditional viscometers are not designed for nano-suspension and hence the measured values of viscosity with them may not give a correct picture. Also during dynamic flow situation, particle collisions give a collision viscosity which cannot be measured by these apparatuses.

B. Experimental Works on Convection in Nanofluids

The first work on convective flow and heat transfer of nanofluids inside 10.66-mm-diameter tube was presented by Pak and Cho [137]. The first observation they made was a substantial increase in heat transfer coefficient in the turbulent flow regime. The increase was 45% with 1.34% Al_2O_3 particles and 75% with 2.78% same particles. One can readily find that this increase is more than the increase in conductivity alone and hence the enhanced convective heat transfer cannot be attributed to the increase in conductivity of the nanofluid alone. They also proposed a modified Dittus–Boelter correlation [141] for their data after conducting regression analysis to their experimental data as

$$Nu = 0.021 Re^{0.8} Pr^{0.5} \tag{24}$$

However, it must be understood that this equation is not of much practical significance as it does not account for particle volume fraction or particle size separately (but through Re, nanofluid conductivity, and Pr). They found that the Darcy friction factor follows the Kays correlation. Thus, due to rise in viscosity, there is a substantial rise in frictional pressure drop. This means that although the heat transfer coefficient increases in nanofluids, the penalty of pressure drop is substantial.

In convective heat transfer applications, there is always a competition between the enhancement of heat transfer and the resulting pressure penalty. Any enhancement method such as creating high turbulence and interruption of boundary layer has a pressure penalty associated with it and this result in higher pumping power requirement which may offset the advantage of heat transfer enhancement. Often we compare heat transfer enhancement at the same pumping power which gives a better picture. In case of nanofluids, Pak and Cho [137] claimed that even though the Nusselt number increased for the dispersed fluid with increasing volume concentration and Re, the convective heat transfer coefficient of the dispersed fluid was found to be 12% lower than that of pure fluid when compared at the same average velocity.

Xuan and Li [142] came out with experimental results which gave a bright perspective for nanofluid. One major difference with that of Pak and Cho [137] was that their particles were of pure copper particles and of slightly larger size (~100 nm). On the contrary to Pak and Cho [137], Xuan and Li [142] showed an increase of as high as 40% in heat transfer coefficient at a given velocity. They explained it by stating that in the study of Pak and Cho [137], the increase in viscosity was large which might have suppressed the turbulence reducing the heat transfer. Hence, they indicated that not only the volume fraction is important but also the particle dimensions and material properties are important and if the fluid is designed properly, substantial rise in heat transfer coefficient is on the card.

The results clearly indicate that the Dittus–Boelter equation [141] with modified nanofluid properties is not enough to describe convection in nanofluids. In other words, a nanofluid cannot be treated as a single fluid just by changing properties to effective properties. In the convection of nanofluid, there are distinct additional effects such as gravity, Brownian force, drag on the particle, and diffusion. The discrepancy between the Difttus–Boelter equation and the nanofluid convection was found to be 39% with 2% Cu particles. Following the theory of thermal dispersion, they suggested a correlation for turbulent nanofluid heat transfer inside a pipe in the form:

Laminar flow:
$$\mathrm{Nu} = 0.4328(1 + 11.285\phi_p^{0.754}\mathrm{Pe}_d^{0.218})\mathrm{Re}^{0.333}\mathrm{Pr}^{0.4} \quad (25)$$

Turbulent flow:
$$\mathrm{Nu} = 0.0059(1 + 7.6286\phi_p^{0.6886}\mathrm{Pe}_d^{0.001})\mathrm{Re}^{0.9238}\mathrm{Pr}^{0.4} \quad (26)$$

where,

Pe_d is the particle Peclet number, $\dfrac{u_p d_p}{\alpha_p}$;

α_p is the thermal diffusivity of the particle; and

ϕ_p is the particle volume fraction.

Another important investigation on convective heat transfer in nanofluids was by Wen and Ding [143]. This work showed for the first time the effect of entrance length in a flow inside a tube. Laminar flows usually have long hydrodynamic and thermal entrance regions. In these regions since the boundary layer is thinner, the heat transfer coefficient is higher. Wen and Ding [143] presented the measurement of local heat transfer coefficient along the tube during laminar flow as shown in Fig. 24. They used different

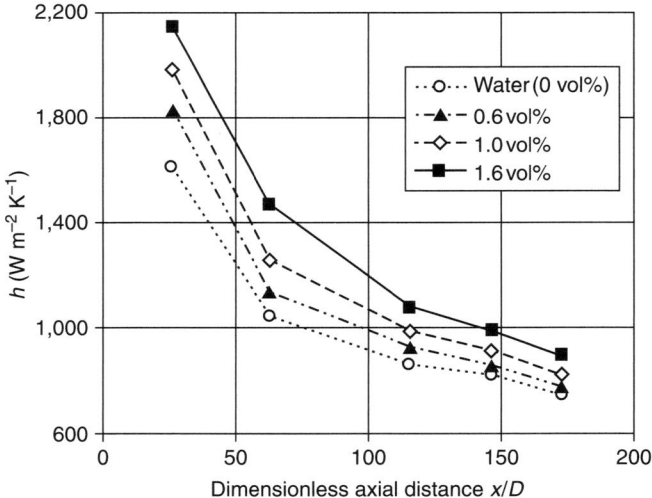

FIG. 24. The measured local heat transfer coefficient for convection inside a tube presented by Wen and Ding [143].

concentrations of water–γ-Al$_2$O$_3$ combination flowing inside a copper tube of 4.5 mm inner diameter having a length of 970 mm, which was electrically heated from outside giving a constant heat flux boundary condition.

For data reduction, they estimated viscosity by the Einstein's Eq. (22). The results showed a substantial rise in heat transfer coefficient. The most interesting feature that can be observed here is the fact that the increase in heat transfer coefficient is more at the entry length region and the enhancement increases with particle concentration. This not only tells about a steady entrance region effect but also talks about having higher enhancement of heat transfer by "smart" options like interruption of boundary layer and creating artificial entrance region. For comparison with existing heat transfer correlation, they used Shah's [144] correlation for thermal entry length with constant heat flow in the form:

$$\text{Nu} = 1.953 \left(\text{RePr} \frac{D}{x} \right)^{1/3} \quad \text{for} \quad \left(\text{RePr} \frac{D}{x} \right) \geq 33.3 \qquad (27)$$

$$\text{Nu} = 4.364 + 0.0722 \left(\text{RePr} \frac{D}{x} \right) \quad \text{for} \quad \left(\text{RePr} \frac{D}{x} \right) < 33.3 \qquad (28)$$

The experimental results varied much from that of the Shah's [144] equation, which gives an idea that apart from the regular property enhancement some special effects are also playing an account. Yet another important observation of Wen and Ding [143] was that the entry length of nanofluids is longer than the pure base fluid. While commenting on the reason for the enhancement and the special effects involved, Wen and Ding [143] indicated few possibilities. They pointed out that particle migration and nonuniform distribution of thermal conductivity and viscosity may lead to reduction of the boundary layer thickness with increasing heat transfer. However, this proposition was just a suggestion and was not conclusively proven by their study.

Yang et al. [145] presented results of their experiments on a similar test rig as that of the previous investigators. They used tubes of 4.57 mm inner diameter and 457 mm (i.e., 100 diameters) long. One important feature of their test loop was its small volume of hold-up fluid and the other was using hot water for heating instead of electrical heating. This second feature may be an important one because Kabelac [146] has recently indicated that electrical heating may affect particle movements in nanofluids in which the particles are likely to carry electrical charge. They used four different experimental fluids of different combinations of two base fluids (one is an automatic transmission fluid and the other is a mixture of synthetic oils with additives) and graphite particles between 2 and 2.5% concentration. The particles were disc shaped with 1–2 nm diameter and 20–40 nm thickness.

Their conclusion was that the particles loading, temperature, source of the nanoparticles, and base fluid used affect the heat transfer results. The results were plotted keeping Sieder Tate [147] correlation for laminar developing flow as datum in the form:

$$\Omega = 1.86 \, Re^{1/3} \qquad (29)$$

where $\Omega = Nu \, Pr^{-1/3} (L/D)^{1/3} (\mu_b/\mu_w)^{-0.14}$

Although overall results are found to agree with this, the scatter is large at the lower Reynolds number. A comparison of the experimental data with the above equation is shown in Table II. This shows a much lower amount of enhancement in comparison with the data of Yang et al. [145] and Sieder-tate [147] correlation.

With their experiment, they then predicted values of the constants of the equation. They also compared their results with two other correlations. One is Oliver [148] correlation.

$$Nu \left(\frac{\mu_w}{\mu_b}\right)^{0.4} = 1.75 \left(Gz + 5.6 \times 10^{-4} \left(Gr \, Pr \frac{L}{D}\right)^{0.7}\right)^{1/3} \qquad (30)$$

where

Gz = Graetz number = $(\dot{m}C_p)/kL$ and
Gr = Grashof number = $(\rho_f^2 \beta_f g \Delta T D^3)/\mu^2$.

The other correlation is that of Eubank and Proctor [149] for laminar flow in horizontal tube.

TABLE II

HEAT TRANSFER COEFFICIENT RATIOS OF NANOFLUID VERSUS CORRESPONDING BASE FLUID

	Heating fluid temperature (°C)	Eq.(29)	Experiment
h(EF#1.1)/h(BF#1)	50	1.19	1.03
	70	1.19	1.03
h(EF#1.1)/h(BF#1)	50	1.36	1.22
	70	1.36	1.15
h(EF#1.1)/h(BF#1)	50	1.02	1.01
	70	1.02	1.01
h(EF#1.1)/h(BF#1)	50	1.14	1.08
	70	1.14	1.07

$$\text{Nu}\left(\frac{\mu_w}{\mu_p}\right)^{0.14} = 1.75\left(\text{Gz} + 12.6\left(\text{Gr}\,\text{Pr}\,\frac{D}{L}\right)^{0.4}\right)^{1/3} \tag{31}$$

The results fall in between them. However, they failed to indicate that in reality the slope of the experimental curve is higher than that suggested by both the equations, indicating a much higher entry length effect than these equations suggest. This work by Yang et al. [145] has got several deviations from other data. This may be due to the fact that the particles are disc type and the major dimension (diameter) is too large to be called nanoparticles. Hence, there is a doubt whether this work falls in category of nanofluids or not.

Anther work by Heris et al. [150] in recent times brought out conclusion which is similar to that by Xuan and Li [142]. Here, the test was conducted in a 6-mm-diameter copper tube for water–CuO and water–Al_2O_3 nanofluids. Substantial enhancement was reported with higher enhancement for Al_2O_3-based nanofluid. It was found that Sieder–Tate [147] correlation for turbulent flow is inadequate to predict the heat transfer enhancement with these nanofluids. The two important features of this work were, first the observation that the heat transfer enhancement increases significantly with particle volume fraction and, second, the enhancement is more at higher Peclet number. Figure 25 shows the above effect.

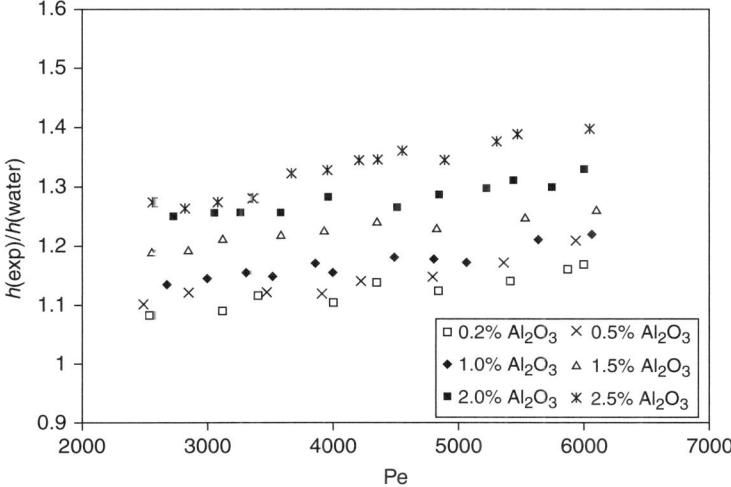

FIG. 25. The increase of heat transfer coefficient of water–Al_2O_3 nanofluids against Peclet number by Heris et al. [150].

Their explanation about this increase was similar to that of Xuan and Li [142], namely dispersion, chaotic particle movement, and Brownian motion. These results are significant on many counts. First, here the authors carried out experiments with steam as the heating medium and the problems envisaged by Kabelac [146] regarding electrical effect on the particles should not be present. Second, unlike Xuan and Li [142], nonmetallic particles were used here and still the enhancement was high enough to be marked and traditional turbulent flow equations seem to be inadequate to describe it. In general, it seems that particle source, method of preparation, technique of dispersion, size distribution, pH value, and a large number of other issues are responsible for these divergent trends in experimental data between Pak and Cho [137] and Yang et al. [145] on one hand and Xuan and Li [142], Wen and Ding [143], and Heris et al. [150] on the other.

Finally, one interesting experimental work came out from Ding et al. [140] on the convection of CNT-containing nanofluids. They used MWCNT. Probably, this is the only study on convection with CNT-containing nanofluids. Since CNTs have a tendency to agglomerate, they used high-speed (24,000 rpm) rotor to disperse it properly. They first measured the thermal conductivity of the nanofluids which showed upto 50% of enhancement with 0.7% CNT. It was also interesting to note that there was also a tremendous temperature effect on conductivity as shown in their results (Fig. 26) over just 10% rise in suspension temperature.

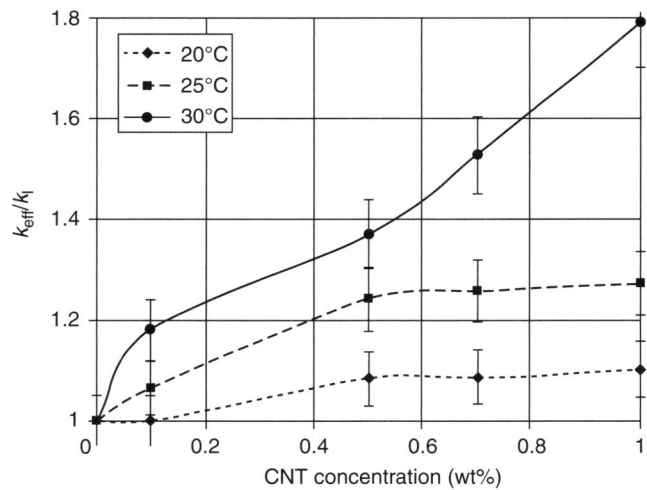

FIG. 26. Measured thermal conductivity of CNT nanofluids by Ding et al. [140].

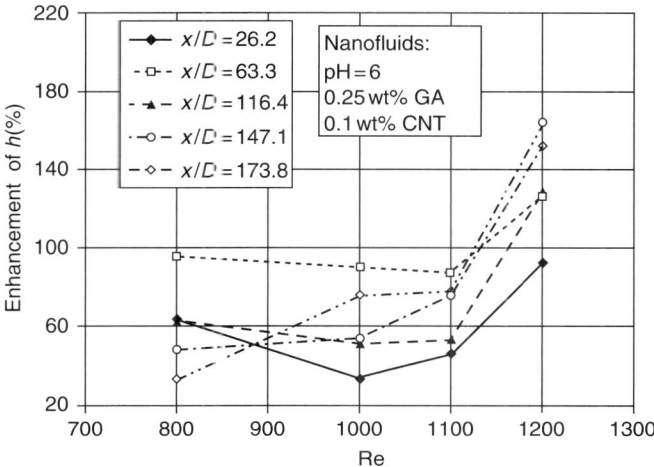

FIG. 27. Effect of Reynolds number on the heat transfer coefficient of water–CNT nanofluids by Ding et al. [140].

The convective heat transfer of CNT nanofluids showed considerable improvement as compared with that of nanoparticle-suspended nanofluids. Since in this study gum arabic was used as the stabilizing agent, the studies were always compared with water having gum Arabic alone (0% CNT). The enhancement was tested against parameters such as particle concentration, axial distance, Reynolds number, and pH value. Figure 27 shows one such effect.

As much as 350% enhancement of heat transfer coefficient was observed at Re = 800. The CNT volume concentration used was small (<0.5%). The entrance length effect was observed, but unlike Wen and Ding [143], the enhancement increased along the entry length, reached a maximum, and then decreased. The point of maximum enhancement was found to increase with Reynolds number and particle concentration. The enhancement increases with Reynolds number and was not much affected by pH value of the suspension. The large amount of enhancement shown in this study cannot be attributed to increased thermal conductivity alone. They found that the enhancement increases suddenly enormously beyond a particular Reynolds number which they attributed to shear thinning. Furthermore, they suggested that particle rearrangement of high aspect ratio (>100) of CNTs and reduction of boundary layer thickness by nanotubes are important additional mechanisms.

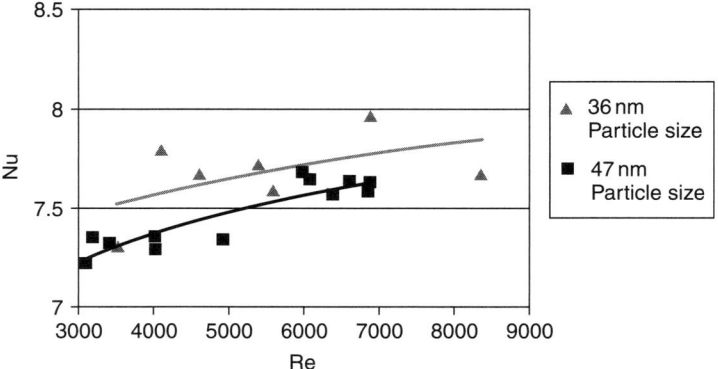

FIG. 28. Particle size effect on convective heat transfer as shown by Nguyen *et al.* [151].

Recently, in an experimental study, Nguyen *et al.* [151] showed the effect of particle size in the convective heat transfer of Al_2O_3 nanoparticle–water mixture for turbulent flow regime. They investigated on the heat transfer enhancement behavior of the nanofluid, flowing inside a closed system destined for cooling of microprocessors or other electronic components. For a particular nanofluid with 6.8% particle volume concentration, heat transfer coefficient has been found to increase as much as 40% compared to that of the base fluid. It has also been found that an increase of particle concentration has produced a clear decrease of the heated component temperature. Experimental data have clearly shown that nanofluid with 36 nm particle diameter provides higher heat transfer coefficients than the nanofluid with 47 nm particle size. A typical result is depicted in Fig. 28.

C. Mechanisms in Convection of Suspensions

Making a proper numerical model for convective heat transfer requires systematic understanding of the basic mechanisms involved in it. While good amount of research in the past gave a clear idea of the hydrodynamic and thermal mechanisms involved in suspensions with macro- and micro-sized particles, applying those mechanisms to nanoparticle suspension requires more methodical experimental studies. Thus the mathematical treatment of these nanoparticle suspension flow and heat transfer behavior is considerably complex; there is no guarantee that these methods used in suspensions will work in the same form with nanofluids. However, these methods of modeling as well as the resulting heat transfer correlations can be a very good starting point for the analytical and experimental treatment of convection in nanofluids.

To ascertain the hydrodynamics of suspensions, one has to first choose the frame of reference for the analysis. This frame can be Eulerian (static frame) or Lagrangian (frame moving with the particle) approach. However, it is always convenient to model the fluid flow in Eulerian frame.

The popular one in the Eulerian approach is the single fluid model. Here the entire suspension is treated as a single fluid whose properties are in between the solid and fluid properties which form the suspensions. Defining these properties is often tricky and may need experimental input. Hence, the continuity, momentum, and energy equations can be written. The continuity equation is

$$\frac{\partial}{\partial t}(\rho_m) + \frac{\partial(\rho_m u_m)}{\partial x} + \frac{\partial(\rho_m v_m)}{\partial y} + \frac{\partial(\rho_m w_m)}{\partial z} = 0 \qquad (32)$$

Here, u_m, v_m, and w_m are mass-averaged velocities of the single fluid in x, y, and z directions, respectively. ρ_m is the average density of the medium. For example, u_m is given by

$$u_m = \frac{\phi_p \rho_p u_p + \phi_f \rho_f u_f}{\rho_m} \qquad (33)$$

where the quantities ϕ_p, ρ_p, and u_p are the volume fraction, density, and velocity of the particle; and ϕ_f, ρ_f, and u_f are the same for the fluid. The average density of the medium (or assumed single fluid) is given by

$$\rho_m = \phi_p \rho_p + \phi_f \rho_f \qquad (34)$$

The momentum equation for incompressible suspension (solid in liquid) is given by

$$\frac{\partial}{\partial t}(\rho_m u_m) + u_m \frac{\partial u_m}{\partial x} + v_m \frac{\partial u_m}{\partial y} + w_m \frac{\partial u_m}{\partial z} = -\frac{\partial P}{\partial x}$$
$$+ (\mu_m + \mu_t)\left(\frac{\partial^2 u_m}{\partial x^2} + \frac{\partial^2 u_m}{\partial y^2} + \frac{\partial^2 u_m}{\partial z^2}\right) \qquad (35)$$
$$+ \rho_m g + F_x + \frac{\partial}{\partial x}(\phi_p \rho u^2_{pd})$$

where u_{pd} is particle drift velocity given by

$$u_{pd} = u_p - u_m$$

and μ_t is the turbulent viscosity of the suspension. The medium viscosity is to be taken from experimental data on the viscosity of the suspension. For the evaluation of turbulent viscosity, we need a turbulence model. In recent times, more popular had been the "k–ε model" where additional transport equations for turbulent kinetic energy (k) and its dissipation rate (ε) are solved and finally the turbulent viscosity is calculated as

$$\mu_t = \rho C_\mu \frac{k^2}{\varepsilon} \tag{36}$$

where C_μ is a constant (usually ~ 0.09).

As far as the viscosity is concerned, it is difficult to model in a particulate flow. Often it is taken as the sum of collisional viscosity, kinetic viscosity, and frictional viscosity, and each of these terms is evaluated by empirical correlations available in literature. It seems that for nanofluids due to its stable nature the best possible option is to use experimental value. For dilute suspensions ($\phi_p < 2\%$), the famous Stoke–Einstein formula for viscosity can be a good approximation for nanofluids.

$$\mu = \mu_L(1 + 2.5\phi_p) \tag{37}$$

In a similar way, the momentum equation for y and z direction can be written.

The drift velocity can be correlated to interfacial drag and particle acceleration. In case there is no slip between the particle and the fluid (i.e., the particle moves with the fluid), the last term of Eq. (35) drops out.

The energy equation for this case is

$$\frac{\partial}{\partial t}(\phi_p \rho_p h_p + \phi_f \rho_f h_f) + \phi_p \rho_p \left(u_p \frac{\partial h_p}{\partial x} + v_p \frac{\partial h_p}{\partial y} + w_p \frac{\partial h_p}{\partial z} \right) \\ + \phi_f \rho_f \left(u_f \frac{\partial h_f}{\partial x} + v_f \frac{\partial h_f}{\partial y} + w_f \frac{\partial h_f}{\partial z} \right) = k_{\text{eff}} \left(\frac{\partial^2 T}{\partial x^2} + \frac{\partial^2 T}{\partial y^2} + \frac{\partial^2 T}{\partial z^2} \right) \tag{38}$$

Here, h is the enthalpy of the corresponding phase which can be replaced by $C_p T$ (where C_p is the specific heat) and k_{eff} is the effective conductivity of the medium.

It should be mentioned here that due to interaction of the particles and particle fluid, in single fluid formulation a thermal dispersion term is usually added to the energy equation which acts as an additional virtual conduction. The single fluid equation with dispersion term can be written as

$$(\rho C_p)_m \left[\frac{\partial T_m}{\partial t} + u_m \frac{\partial T_m}{\partial x} + v_m \frac{\partial T_m}{\partial y} + w_m \frac{\partial T_m}{\partial z} \right] = (k_{eff} + k_d) \left(\frac{\partial^2 T_m}{\partial x^2} + \frac{\partial^2 T_m}{\partial y^2} + \frac{\partial^2 T_m}{\partial z^2} \right) \quad (39)$$

Here, the temperature T_m is the mean medium temperature (which is assumed to be in thermal equilibrium between particle and fluid). k_{eff} is the effective-medium conductivity and k_d is the dispersive equivalent conductivity of the medium.

In Eulerian–Lagrangian model, the particle is tracked in the Lagrangian frame and the fluid in the Eulerian frame. The Lagrangian particle n momentum equation can be written as

$$\frac{du_p}{dt} = F_D(u - u_p) + \frac{g_x(\rho_p - \rho)}{\rho_p} + F_x \quad (40)$$

where F_D is the drag force given by

$$F_D = \frac{18\mu}{\rho_p d_p^2} \frac{C_D Re}{24} \quad (41)$$

Here, u_p is the particle velocity; u is the fluid velocity; g_x is the component of g in x direction; and d_p particle diameter. Re is the relative Reynolds number given by

$$Re = \frac{\rho d_p |u_p - u|}{\mu}$$

There are various models of the drag coefficient C_D for nanoparticles; the Stokes law seems to be appropriate.

In this model, the particle energy balance equation may be formulated as

$$hA_p(T - T_p) = \rho_p V_p C_p \frac{dT_p}{dt} \quad (42)$$

where A_p is the particle surface area and V_p is the particle volume. The interfacial heat transfer equation is given in terms of Nusselt number. Often following equations are used.

$$\mathrm{Nu} = 1 + (1 + \mathrm{Re_p Pr})^{1/3} \quad \text{for} \quad 0 \leq \mathrm{Re_p} \leq 1 \tag{43}$$

$$\mathrm{Nu} = \mathrm{Re_p}^{0.41} \left(1 + \frac{1}{\mathrm{Re_p Pr}}\right)^{1/3} \mathrm{Pr}^{1/3} + 1 \quad \text{for} \quad 1 \leq \mathrm{Re_p} \leq 100 \tag{44}$$

$$\mathrm{Nu} = 0.752 \, \mathrm{Re_p}^{0.472} \left(1 + \frac{1}{\mathrm{Re_p Pr}}\right)^{1/3} \mathrm{Pr}^{1/3} + 1 \quad \text{for} \quad 100 \leq \mathrm{Re_p} \leq 2000 \tag{45}$$

Here, the particle Reynolds number $\mathrm{Re_p}$ is given by

$$\mathrm{Re_p} = \frac{\rho d_p U}{\mu}$$

where $U = \sqrt{(u - u_p)^2 (v - v_p)^2 + (w - w_p)^2}$

It goes without saying in all the approaches that the final solution requires tedious numerical computer calculations and the chances of getting analytical solutions are very remote. However, for simple cases of single fluid treatment or with large number of assumptions, some analytical solutions are possible but the applicability of these solutions is doubtful. Hence, quite often we may depend on many experimental correlations.

Apart from the effects of fluid–particle interaction, particle–particle collision, and wall particle collision and under certain circumstances, some special effects (forces) become significant in particulate flows. For example, the solid particles suspended in a fluid experience a force in the direction opposite to the imposed temperature gradient. This is called thermophoretic force. Similarly shear lift force, Brownian motion, and Soret and Dufour effects can bring in slip condition to particle laden flows.

D. ANALYTICAL AND NUMERICAL STUDIES ON CONVECTION IN NANOFLUIDS

In this section, analytical and numerical investigation on the heat transfer analysis of nanofluid is made. Major models and methodologies used by different investigators are discussed in detail in this section.

Xuan and Roetzel [152] were probably the first ones to put forward a concept for theorizing convection in nanofluids. Their concept, although abstract and did not have definite proof that such an approach is acceptable, gave a direction in modeling of nanofluid convection. They proposed a dispersion model which takes care of the additional enhancement over effective fluid treatment. This concept originated from the mass dispersion theory of Taylor [153] and Aris [154] which was supported by Dankwert [155]. Later on this concept has been used to porous medium [156] and suspensions [157].

The concept of dispersion is that due to the presence of solid particles the flow of fluid and the transport of heat will not follow the same path as that of the pure fluid. It will be more torturous path and the effect of this can be modeled by adding an equivalent diffusion term in the corresponding energy equation. However, the equivalent thermal conductivity for this modeling is not a real conductivity (or a property of medium) but it is a flow and particle property because this amount of dispersion depends on factors like particle–particle interaction and particle–surface interaction which in turn depends on particle size, particle movement, particle concentration, fluid velocity, and so on. Thus, in essence, we assume that due to all these features there is an additional amount of heat transfer and this additional amount of that transfer corresponds to an additional fictitious conductivity for modeling purpose. This conductivity is known as the thermal dispersion coefficient. Xuan and Roetzel [152] quoted Kaviany [157] to present the basic equations related to thermal dispersion. They considered that the particles induce a velocity and temperature perturbation in the nanofluid given by $\mathbf{u}'$ and T', respectively. Thus,

$$T = \langle T \rangle^f + T' \tag{46}$$

$$u = \langle u \rangle^f + u' \tag{47}$$

Here, the volume-averaged temperature and velocity vectors are

$$\langle T \rangle^f = \frac{1}{V_f} \int_{V_f} T \mathrm{d}V$$

$$\langle \mathbf{u} \rangle^f = \frac{1}{V_f} \int_{V_f} \mathbf{u} \mathrm{d}V$$

Then, the basic convection equation in vector $\mathbf{u}$ can be written for fluid f as

$$\frac{\partial T}{\partial t} + \nabla.\mathbf{u}T = \nabla.(\alpha_f \nabla T) \tag{48}$$

Substituting Eqs. (46) and (47) to this equation and simplifying using the procedure of Kaviany gives

$$(\rho C_p)_{nf} \left[\frac{\partial \langle T \rangle^f}{\partial t} + \langle u \rangle \nabla \langle T \rangle^f \right] = \nabla \left(k_{nf} \nabla \langle T \rangle^f \right) - (\rho C_p)_{nf} \nabla \langle u' T' \rangle^f \qquad (49)$$

Here, nf indicates nanofluid.
In the dispersion model, the additional term in this equation due to perturbations in velocity and temperature the last term is modeled like a conduction flux as

$$(\rho C_p)_{nf} \nabla \langle u' T' \rangle^f = -k_d \nabla \langle T \rangle^f \qquad (50)$$

where k_d is the tensor of dispersive thermal conductivity (also called the dispersion coefficient). This makes Eq. (49) solvable but the dispersion coefficient in k_d needs to be known. As an example of this type of analysis, they considered flow of nanofluid inside a tube which gives the equation.

$$\frac{\partial T}{\partial t} + u \frac{\partial T}{\partial x} = \frac{1}{r} \frac{\partial}{\partial r} \left[\left(\alpha_{nf} + \frac{k_{d,r}}{(\rho C_p)_{nf}} \right) r \frac{\partial T}{\partial r} \right] + \frac{\partial}{\partial x} \left[\alpha_{nf} + \frac{k_{d,x}}{(\rho C_p)_{nf}} \frac{\partial T}{\partial x} \right] \qquad (51)$$

Here, $\langle \ \rangle^f$ notation have been dropped and T and u are used for volume-averaged values. $k_{d,r}$ gives the dispersion coefficient in the radial direction and $k_{d,x}$ that in the axial direction. Now the task remains to determine these dispersion coefficients. Extensive experiments probably can help to find out the dispersion coefficient and the nature of its variation with different parameters such as particle loading, Reynolds number, geometry, and particle size. One such interesting experimental technique may be the one used by Roetzel et al. [158] for determination of dispersion coefficient in a dented tube. The method uses a periodic temperature profile at the entry to the tube which after heat transfer and dispersion gets an amplitude attenuation and phase shift. Since these two measurable quantities are available, the two unknown quantities such as the heat transfer coefficient and thermal dispersion coefficient can be evaluated from it. However, Xuan and Roetzel [152] provided some intuitive suggestion for dispersion coefficient from similar research works as

$$\kappa_d = C(\rho C_p)_{nf} u d_p R \varepsilon_p \quad \text{or} \quad = C^* (\rho C_p)_{nf} u R \qquad (52)$$

where R is the tube radius and C and C^* are constants. These expressions are purely intuitive, and the determination of dispersion coefficient and predicting its value for different situations remains a task to be taken up in future.

Xuan and Roetzel [152] further advanced the concept of using dispersion coefficient by solving Eq. (51) under the assumption that axial dispersion is negligible giving

$$\frac{\partial T}{\partial t} + u\frac{\partial T}{\partial x} = \frac{1}{r}\frac{\partial}{\partial r}\left[\left(\alpha_{nf} + \frac{k_d}{(\rho C_p)_{nf}}\right)\cdot\frac{\partial T}{\partial r}\right] \quad (53)$$

Considering constant inlet and wall temperature, the boundary conditions are given by

$$T|_{r=R} = T_w$$
$$T|_{x=0} = T_0 \quad (54)$$

considering laminar fully developed flow, this equation can be solved by separation of variables to give

$$\frac{T-T_w}{T_0-T_w} = 2\sum_{n=1}^{\alpha} e^{-\beta_m^2 \bar{x}/\overline{Pe}}\frac{J_0(\beta_m \bar{r})}{J_1(\beta_m)\beta_m} \quad (55)$$

where

$$\bar{r} = \frac{r}{R}$$

$$\bar{x} = \frac{x}{L}$$

$$Nu = \frac{h(2R)}{k_{nf}^*}$$

$$Pe^* = \frac{uL}{\alpha_{eff}^*} \quad \text{and} \quad \overline{Pe} = Pe^*\left(\frac{R}{L}\right)^2$$

Here, $k_{eff}^* = k_{nf} + k_d$
β_ms are the positive roots of the equation.

$$J_0(\beta_m) = 0$$

From this temperature profile, the Nusselt number can be deduced as

$$\mathrm{Nu} = \frac{\sum_{m=1}^{\alpha} e^{-\beta_m^2 \bar{x}/\overline{\mathrm{Pe}}}}{\sum_{m=1}^{\alpha} e^{-\beta_m^2 \bar{x}/\overline{\mathrm{Pe}}}/\beta_m^2} \tag{56}$$

Although this expression is similar to the solution of pure fluid, here the conductivity used for Nu is the sum of effective nanofluid conductivity and the dispersive conductivity. If the axial dispersion is not neglected, we need an axial boundary condition. If we assure that the thermal dispersion begins at the entry, for dispersive flow we get a temperature drop at the entry. This is due to the well-known Dankwert's [155] boundary condition which can be expressed as

$$-k_{\mathrm{eff}}^* \frac{\partial T}{\partial x} = uA\rho C_{\mathrm{p}}(T_{\mathrm{o}} - T) \quad \text{at} \quad x = 0 \tag{57}$$

At the exit of the tube, usual derivative boundary condition was used.

$$\frac{\partial T}{\partial x} = 0 \quad \text{at} \quad x = L \tag{58}$$

Under these conditions, the Nusselt number comes out as

$$\mathrm{Nu} = \frac{\sum_{m=1}^{\alpha} X(\bar{x})/[X(0) - X'(0)/\mathrm{Pe}^*]}{\sum_{m=1}^{\alpha} X(\bar{x})/[X(0) - X'(0)/\mathrm{Pe}^*]/\beta_m^2} \tag{59}$$

where

$$X(\bar{x}) = m_2 e^{m_2 + m_1 \bar{x}} - m_1 e^{m_1 + m_2 \bar{x}}$$

$$X'(\bar{x}) = m_1 m_2 (e^{m_2 + m_1 \bar{x}} - e^{m_1 + m_2 \bar{x}})$$

$$m_{1,2} = \frac{\mathrm{Pe}^* \pm \sqrt{\mathrm{Pe}^{*2} + 4\beta_m(L/R)^2}}{2}$$

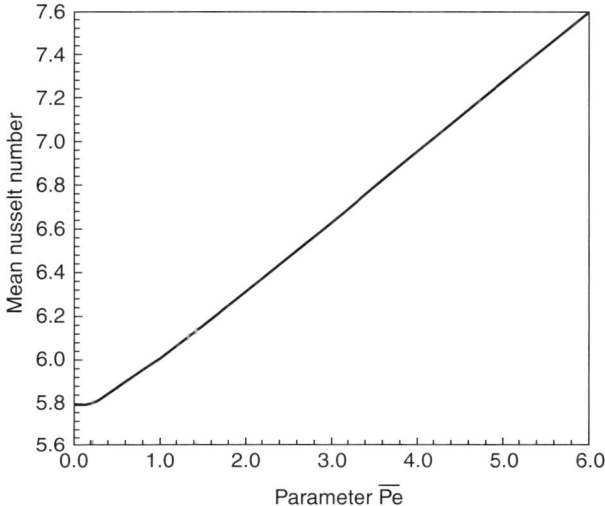

FIG. 29. Nusselt number according to Eq. (28) [152].

This solution was plotted by Xuan and Roetzel [152] as shown in Fig. 29. Using the expression for dispersion coefficient given by Beckaman et al. [159],

$$\frac{k_d}{(\rho C_p)} = \begin{cases} 10.1 Ru\sqrt{f/2} + 5.03 Ru & \text{for large temperature gradient} \\ [10.1 Ru\sqrt{f/2} + 5.03 Ru]/(1 + 1/\sqrt{2f}) & \text{for adiabatic cases} \end{cases} \quad (60)$$

where f is the friction factor.

They indicated that the dispersion coefficient is the effective embodiment of effects such as Brownian diffusion, sedimentation, and dispersion.

While dispersion model tells about an overall modeling strategy, it is important to identify the mechanisms which are probably responsible behind the dispersive behaviors. While it is fact that the true contribution of these mechanisms can only be revealed by highly sophisticated experiments which have not yet been conducted, analytical treatments taking these effects into consideration and comparison of the results with available data can give some indication about the importance of these mechanisms. Among these mechanisms, the particle migration seems to be the most logical one and is likely to play an important role in the convection of nanofluids.

Ding and Wen [160] investigated the particle migration effect in nanofluids. They used a well-known "mass balance approach," the constitutive equation for which was laid by Phillips et al. [161]. The basic concept of

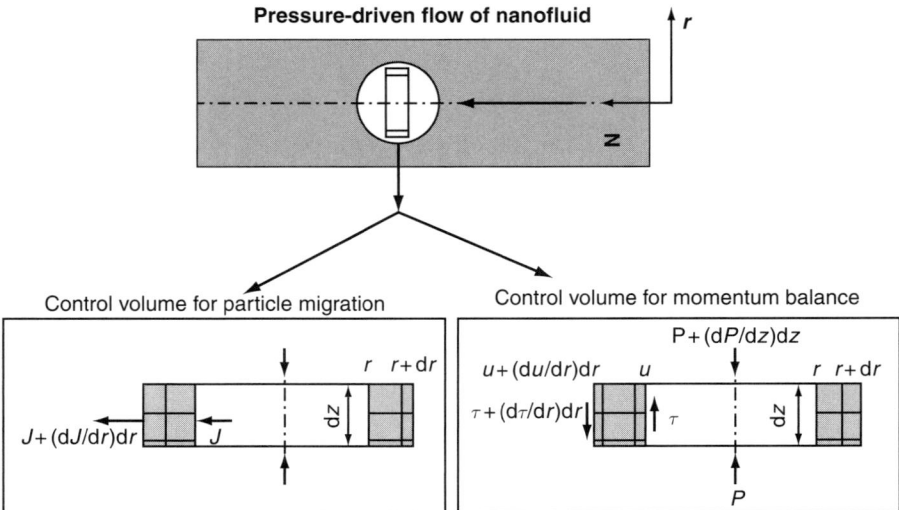

Fig. 30. Control volume considered by Ding and Wen [160] mass flux equation.

this model is that the particles migrate here under the action of shear force from a region of higher shear to a region of lower shear under higher viscosity to lower viscosity and due to Brownian diffusion from the region of higher particle concentration to lower particle concentration. Now considering mass balance over a control volume given in Fig. 30, one can obtain

$$J + r \frac{dJ}{dr} = 0 \qquad (61)$$

where J is the total particle flux in r (radial) direction.

It may be mentioned here in the above equation the particle phase is taken to be continuous. Now this particle migration flux consists of these components, arising out of three mechanisms considered here.

$$J = J_\mu + J_b + J_c \qquad (62)$$

where

$$J_\mu = \text{flux due to viscosity gradient} = -K_\mu \dot{\gamma} \varepsilon_p^2 \left(\frac{d_p^2}{\mu}\right) \frac{d\mu}{d\varepsilon_p} \nabla \varepsilon_p$$

$$J_b = \text{flux due to nonuniform shear} = -K_c d_p^2 (\varepsilon_p^2 \nabla \dot\gamma + \varepsilon_p \dot\gamma \nabla \varepsilon_p)$$

$$J_c = \text{flux due to Brownian motion} = -D_b \nabla \varepsilon_p$$

where

K_μ and K_c are constants;
$\dot\gamma$ = shear rate;
μ = viscosity;
d_p = practical diameter; and
D_b is the Brownian diffusion coefficient given by

$$D_b = \frac{k_b T}{3\pi \mu d_p} \tag{63}$$

where k_b is the Boltzman constant.

Integration of Eq. (62) and imposition of symmetric boundary condition ($J=0$ at $r=0$) yields the following equations for one-dimensional case.

$$K_\mu \dot\gamma \phi_p^2 \frac{d_p^2}{\mu} \frac{d\mu}{dr} + K_s d_p^2 \phi_p^2 \frac{d\dot\gamma}{dr} + K_c d_p^2 \phi_p \dot\gamma \frac{d\varepsilon_p}{dr} + D_b \frac{d\phi_p}{dr} = 0 \tag{64}$$

In the same Fig 30, the momentum balance in the control volume is shown which yields the equation.

$$\frac{1}{r}\left(\frac{dr\tau}{dr}\right) = -\frac{dp}{dz} \tag{65}$$

where p is the pressure, z is the axial coordinate, and τ is the shear stress. Due to symmetry, the shear stress is zero at the axis ($\tau=0$ at $r=0$), giving the solution.

$$\tau = -\frac{r}{2}\left(\frac{dp}{dz}\right) \tag{66}$$

Now to get explicit solution to this, one must know the correlation for shear stress which depends on the rheology of the fluid. Since most of the investigations suggest a Newtonian behavior of nanofluids, a linear correlation between shear stress and shear rate can be assumed as,

$$\tau = -\mu \dot\gamma \tag{67}$$

This reduces the momentum balance to

$$\dot{\gamma} = \frac{1}{2\mu}\left(\frac{dp}{dz}\right) r \tag{68}$$

where the shear rate is given by

$$\dot{\gamma} = \frac{du}{dr}$$

A model has to be used for viscosity. In this case, Ding and Wen [160] used the Batchelor's formula (Eq. 22) given earlier. Now the equations can be nondimensionalized as

$$\bar{\dot{\gamma}} = -\frac{\bar{r}}{\bar{\mu}} = \frac{d\bar{u}}{d\bar{r}} \tag{69}$$

$$\frac{1}{\bar{\mu}}\frac{d\bar{\mu}}{d\bar{r}} + \left(\frac{K_c}{K_\mu}\right)\frac{1}{\bar{\dot{\gamma}}}\frac{d\bar{\dot{\gamma}}}{d\bar{r}} + \left(\frac{K_c}{K_\mu}\right)\frac{1}{\phi_p}\frac{d\phi_p}{d\bar{r}} - \frac{1}{K_\mu \text{Pe}}\frac{1}{\phi_p^2 \bar{r}}\frac{d\phi_p}{d\bar{r}} \tag{70}$$

where

$$\bar{\mu} = \frac{\mu}{\mu_f}$$

$$\bar{\dot{\gamma}} = \dot{\gamma}\frac{2\mu_f}{(dp/dz)R}$$

$$\bar{r} = \frac{r}{R}$$

$$\bar{u} = \frac{2\mu_f}{(dp/dz)R^2} u$$

$$\text{Pe} = \frac{3\pi d_p^3 (-dp/dz)R}{2k_B T}$$

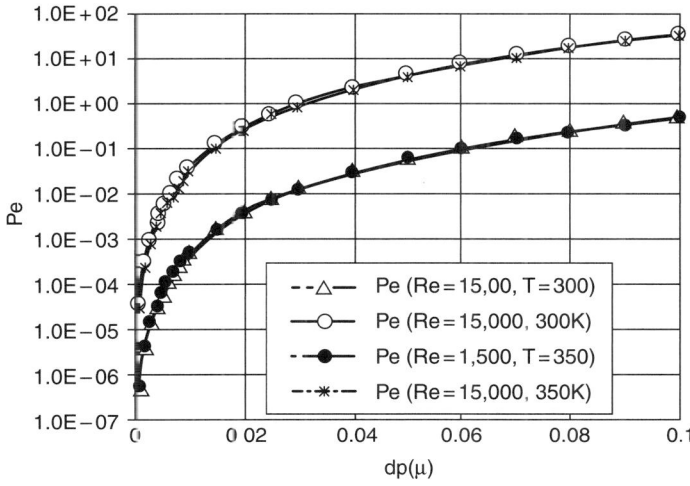

FIG. 31. Effect of temperature and particle size on the particle Peclet number [160].

Solving the above equations, particle concentration profiles across the fluid can be obtained. The key parameter in this equation is the particle Peclet number, Pe. For 90–500 nm particles, the variation of Pe with particle diameter, d_p, is given in Fig. 31

The distribution of particles in the radial direction for different Peclet numbers is given in Fig. 32

It is clear that the higher the Peclet number the more the variation in particle concentration. This result also indicates that there is a possibility of agglomeration at the core region due to high particle concentration which is unlikely to be dispersed by the shear due to the low magnitude of shear at the core region. However, these results may get changed when shear thinning behavior is considered due to non-Newtonian behavior.

The other interesting result was the large variation of viscosity from the core to the wall due to variation in particle concentration. It has to be kept in mind that this variation of viscosity as shown in Fig. 33 is not due to shear thinning (which is not considered here) but due to particle migration.

This gives reason for why nanofluids are likely to give higher heat transfer with relatively lower pressure penalty due to lower viscosity near the wall. They also showed that the particle distribution is more nonuniform for higher particle concentrations. This also indicates that in the analysis of thermal transport in nanofluids one must be concerned that the near wall region may have lower particle concentration, leading to lower thermal conductivity near the wall and lower heat transfer (also lower shear stress).

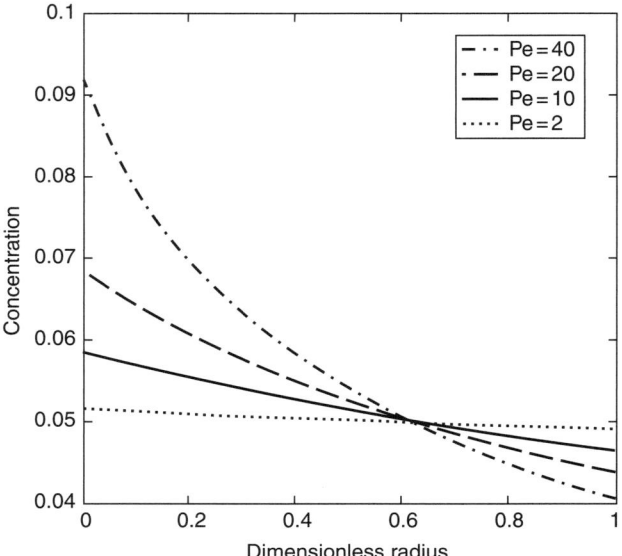

FIG. 32. Influence of the Peclet number on particle distribution [160].

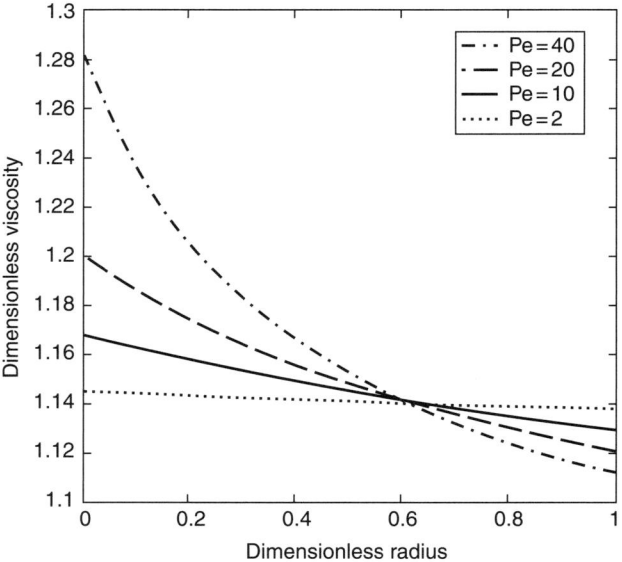

FIG. 33. Influence of the Peclet number on viscosity distribution [160].

Hence, Ding and Wen [160] predicted that there may be an optimum particle size for a compromise between heat transfer and pressure drop.

Buongiorno [162] conducted an exhaustive analysis on the slip mechanisms involved in the nanofluid convective flow, which gave a major insight into the particle–fluid interaction. All possible mechanisms of fluid–particle slip during convection of nanofluids, namely inertia, Brownian diffusion, thermophoresis, diffusiophoresis, Magnus effect, fluid drainage, and gravity effects, were discussed. By comparing all the timescales of the above processes, he concluded that for laminar flow (also in the viscous sublayer of the turbulent flow) thermophoresis and Brownian diffusion are important mechanisms while in the turbulent region the nanoparticles are carried by turbulent eddies without any slip and the above diffusion mechanisms are negligible there. Based on these assumptions the continuity equation for nanofluids and nanoparticles were derived in the form:

$$\nabla . \vec{u} = 0 \tag{71}$$

$$\frac{\partial \phi_p}{\partial t} + \vec{v} . \nabla \phi_p = \nabla . \left[D_B \nabla \phi_p + D_T \frac{\nabla T}{T} \right] \tag{72}$$

The momentum equation was proposed in the line of Bird, Stewart, and Lightfoot [163] as

$$\rho \left[\frac{\partial \vec{v}}{\partial t} + \vec{v} . \nabla \vec{v} \right] = -\nabla P - \nabla . \tau \tag{73}$$

The stress terms under the assumption of Newtonian incompressible flow becomes

$$\tau = -\mu [\nabla \vec{v} + (\nabla \vec{u}^*)'] \tag{74}$$

In nanofluids the viscosity is a function of concentration and hence the above three equations are not independent of each other. The energy equation, under the assumption of the presence of thermophoresis and Brownian diffusion effects, takes the form:

$$\rho C \left(\frac{\partial T}{\partial t} + \vec{u} . \nabla T \right) = \nabla . (k \nabla T) + \rho_p C_p \left[D_B \nabla \varepsilon_p \nabla T + D_T \frac{\nabla T . \nabla T}{T} \right] \tag{75}$$

Here, ρC is for the nanofluid, while $\rho_p C_p$ is for the particle phase. The last term on the right-hand side brings the effect of Brownian diffusion and thermophoresis. These equations can be nondimensionalized as:

$$\nabla \cdot \vec{v} = 0 \tag{76}$$

$$\frac{\partial \phi_p}{\partial \xi} + \vec{v} \cdot \nabla \phi_p = \frac{1}{Re_e Sc} \left[\nabla^2 \phi_p + \frac{\nabla^2 \theta}{N_{BT}} \right] \tag{77}$$

$$\frac{\partial \vec{v}}{\partial \xi} + \vec{v} \nabla \vec{v} = -\nabla \psi + \frac{\nabla^2 \vec{v}}{Re}$$

$$\frac{\partial \theta}{\partial \xi} + \vec{v} \nabla \theta = \frac{1}{Re\, Pr} \left[\nabla^2 \theta + \frac{\nabla \phi_p \cdot \nabla \theta}{Le} + \frac{\nabla \theta \cdot \nabla \theta}{Le N_{BT}} \right] \tag{78}$$

where

$$\vec{v} = \frac{\vec{v}}{\bar{v}}, \quad \phi_p = \frac{\phi_p}{\phi_{pb}}, \quad \theta = \frac{T - T_b}{\Delta T}$$

$$\psi = \frac{P}{\rho \bar{v}^2}, \quad R = \frac{r}{D}, \quad \xi = \frac{t}{(D/\bar{v})}$$

where $\bar{v}, \phi_{pb}, \Delta T$, and D are the reference values for these quantities and the nondimensional numbers.

$$Re = \frac{\rho \bar{v} D}{\mu} = \text{Reynolds number}$$

$$Sc = \frac{\mu}{\mu D_B} = \text{Schmidt number}$$

$$N_{BT} = \frac{\varepsilon_{pb} D_B T_b}{D_T \Delta T} = \frac{\text{Brownian diffusivity}}{\text{Thermophore diffusivity}}$$

$$Pr = \text{Prandtl number}$$

$$Le = \text{Lewis number} = \frac{k}{\rho_p C_p D_B \varepsilon_{pb}}$$

Assuming that the axial transport terms are small compared to radial ones, the turbulent transport equations were derived as:

$$\frac{d}{dy}\left[(D_B + D_p)\frac{d\phi_p}{dy} + \frac{D_T}{T}\frac{dT}{dy}\right] = 0 \qquad (79)$$

$$\frac{d}{dy}\left[\mu + \rho D_M \frac{dv}{dy}\right] = 0 \qquad (80)$$

$$\frac{d}{dy}\left[(k + \rho c D_H)\frac{dT}{dy}\right] = 0 \qquad (81)$$

Here, y is the radial coordinate, and D_p, D_M, and D_H are diffusivities of particle eddy, momentum, and heat in the turbulent sublayer; they assumed

$$D_p \sim D_\mu$$

$$\varepsilon_p \sim \varepsilon_{pb}$$

due to mixing with eddies. Eliminating temperature gradient within the above equations for laminar sublayer, the equations can be solved to get particle concentration distribution as

$$\varepsilon = \varepsilon_b e^{\frac{1}{N_{BT}}\left(1 - \frac{y}{\delta_c}\right)} \qquad (82)$$

The plot of this result is shown in Fig. 34 which shows a distribution quantitatively different from that of Ding and Wang [160] mainly due to the inclusion of thermophoresis and neglecting the migration under shear and viscosity gradient.

The final heat transfer equation was obtained after comparison with Prandtl analogy correlation and Gnielinski correlation as

$$Nu_b = \frac{\frac{f}{8}(Re_b - 1000)Pr_b}{1 + \delta_v^+ \sqrt{\frac{f}{8}}(Pr_v^{2/3} - 1)} \qquad (83)$$

where δ_v^+ is an empirical constant and Pr_b is the Prandtl number evaluated at mean viscous sublayer temperature. The results plotted against the

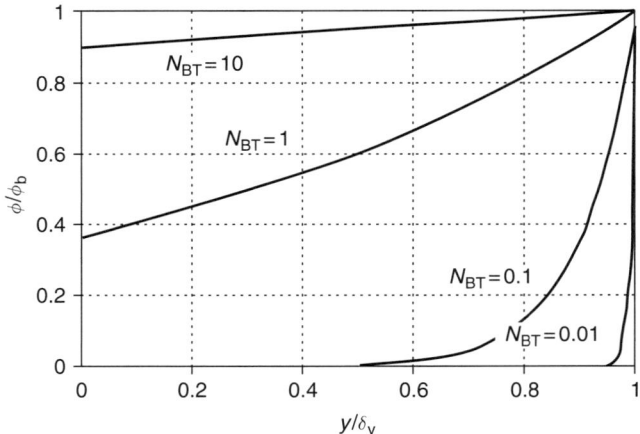

FIG. 34. Particle volume fraction variation presented by Boungiorno [162].

correlation are obtained from the data of Xuan and Li [142] and Pak and Cho [137] along with the well-known Dittus–Boelter equation. The results show that the present model agrees with the Pak and Cho [137] but under predicts the Xuan and Li [142] correlation at $\varepsilon_p > 0$. They attributed this to the temperature effects on thermal conductivity and variation of viscosity near the wall. But this does not explain the entire story of nanofluid convection though it gives important insight.

In the field of numerical modeling of nanofluids, Maiga et al. [164] presented the forced convective heat transfer in nanofluids in two different geometries (uniformly heated tube and radial channel) using finite volume technique. The geometries they considered are shown in Fig. 35.

Their formulation was that of a "single-fluid approach" with fluid properties replaced by the effective nanofluid properties. Hence the basic conservation equations for mass momentum and energy are the usual convection equations given respectively by

$$\nabla \cdot (\rho \vec{V}) = 0 \qquad (84)$$

$$\nabla \cdot (\rho \vec{V} \vec{V}) = -\nabla P + \mu \nabla^2 \vec{V} \qquad (85)$$

$$\nabla \cdot (\rho \vec{V} C_p T) = \nabla \cdot (k \nabla T) \qquad (86)$$

They considered both constant wall temperature and constant wall heat flux at the tube wall as boundary conditions. Symmetry was assumed about

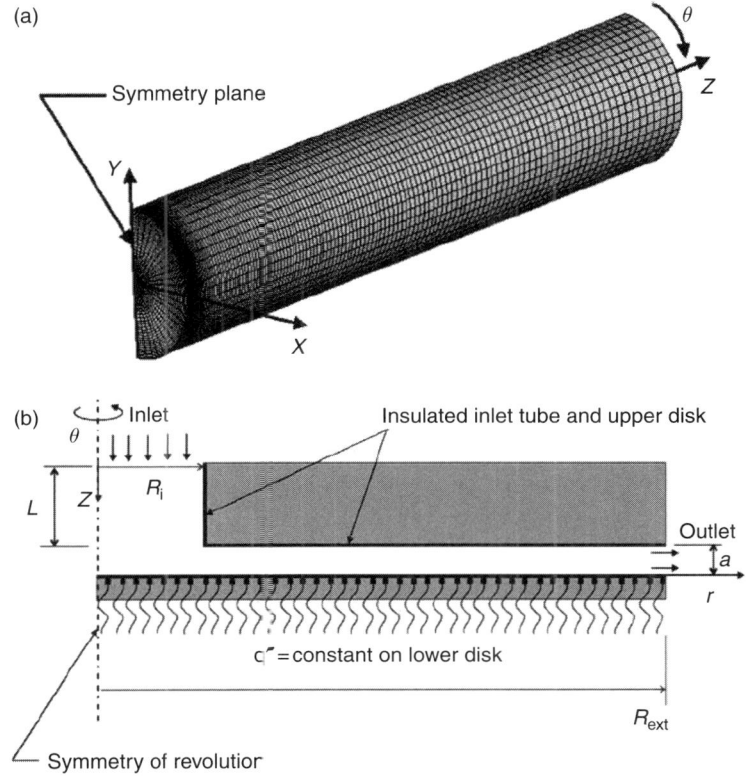

FIG. 35. Geometries studied by Maiga et al. [164]: (a) uniformly heated tube and (b) radial channel between heated discs.

vertical plane through the axis. For the radial channel, a constant heat flux at the impinging wall and insulated upper wall was assumed.

The effective properties of nanofluids for the numerical analysis were taken as below.

$$\rho_{\mathrm{r.f}} = (1 - \phi_{\mathrm{p}})\rho_{\mathrm{bf}} + \phi_{\mathrm{p}}\rho_{\mathrm{p}} \tag{87}$$

$$(C_{\mathrm{p}})_{\mathrm{nf}} = (1 - \phi_{\mathrm{p}})(C_{\mathrm{p}})_{\mathrm{bf}} + \phi_{\mathrm{p}}(C_{\mathrm{p}})_{\mathrm{p}} \tag{88}$$

$$\mu_{\mathrm{nf}} = \mu_{\mathrm{bf}}(123\phi_{\mathrm{p}}^2 + 7.3\phi_{\mathrm{p}} + 1) \quad \text{for water} - \mathrm{Al}_2\mathrm{O}_3 \tag{89}$$

$$= \mu_{bf}(306\phi_p^2 - 0.19\phi_p + 1) \quad \text{for ethylene glycol–Al}_2\text{O}_3 \tag{90}$$

$$k_{nf} = k_{bf}(4.97\phi_p^2 + 2.72\phi_p + 1) \quad \text{for water–Al}_2\text{O}_3 \tag{91}$$

$$= k_{bf}(28.905\phi_p^2 + 2.8273\phi_p + 1) \quad \text{for ethylene glycol–Al}_2\text{O}_3 \tag{92}$$

Here, "bf" stands for base fluid and "nf" nanofluid.

These equations were chosen by fitting curves through regression analysis of experimental data available for nanofluids (particularly for viscosity and thermal conductivity). It was found that in laminar flow region, both water–Al$_2$O$_3$ and ethylene glycol–Al$_2$O$_3$ nano-fluids, the enhancement of average heat transfer coefficient with particle volume fraction was low, while at higher volume fraction the average heat transfer coefficient increased rapidly. One such result is shown in Fig. 36.

Based on their simulation, the following correlations were suggested.

$$\overline{N}_u = 0.086 \, \text{Re}^{0.55} \, \text{Pr}^{0.5} \quad \text{for contact wall heat flux} \tag{93}$$

$$\text{Nu} = 0.28 \, \text{Re}^{0.35} \, \text{Pr}^{0.36} \quad \text{for contact wall temperature} \tag{94}$$

However, they found that the shear stress at wall also significantly increased as shown in Fig. 37.

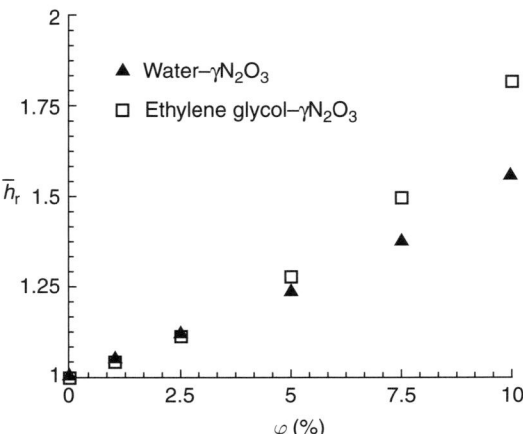

FIG. 36. Influence of Re and φ ($=\varepsilon_p$) on the enhancement of average heat transfer coefficient for laminar convection in nanofluids. [164].

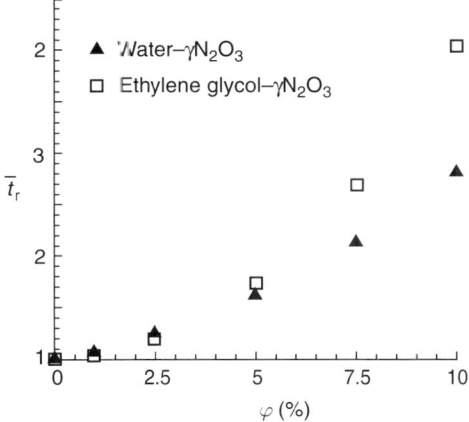

FIG. 37. Increase of shear stress ratio for nanofluids [164].

For the radial flow situation, they found that both the gap between the disc and the Reynolds number have very little effect on the heat transfer enhancement.

It must be kept in mind that the above results are purely numerical and no experimental evidence has been provided in the paper. In view of the fact that it used a "single-fluid model," the results need to be at least qualitatively validated by carefully designed experiment.

The same results with much more elaboration in the discussion of the numerical results were published by Maiga et al. [164] and Roy et al. [165]. Palm et al. [166] extended the radial channel flow problem with the consideration of temperature-dependent properties. They evaluated temperature-dependent properties by fitting curves to the experimental data of Putra et al. [167] for Al_2O_3–water nanofluids in the form:

$$\begin{aligned}\mu_{nf} &= 2.9 \times 10^{-7}T^2 - 2 \times 10^{-4}T + 3.4 \times 10^{-2} \quad \text{at} \quad p = 1\% \\ &= 3.4 \times 10^{-7}T^2 - 2.3 \times 10^{-4}T + 3.9 \times 10^{-2} \quad \text{at} \quad p = 4\%\end{aligned} \quad (95)$$

and

$$\begin{aligned}k_{nf} &= 0.003352T - 0.3708 \ (\text{W m}^{-1}\text{K}^{-1}) \quad \text{at} \quad p = 1\% \\ &= 0.004961T - 0.8078 \ (\text{W m}^{-1}\text{K}^{-1}) \quad \text{at} \quad p = 4\%\end{aligned} \quad (96)$$

They observed that the use of a variable property model predicts higher thermal and hydraulic performances. For example, the local wall temperature was found to reduce when variable property model is used as shown in

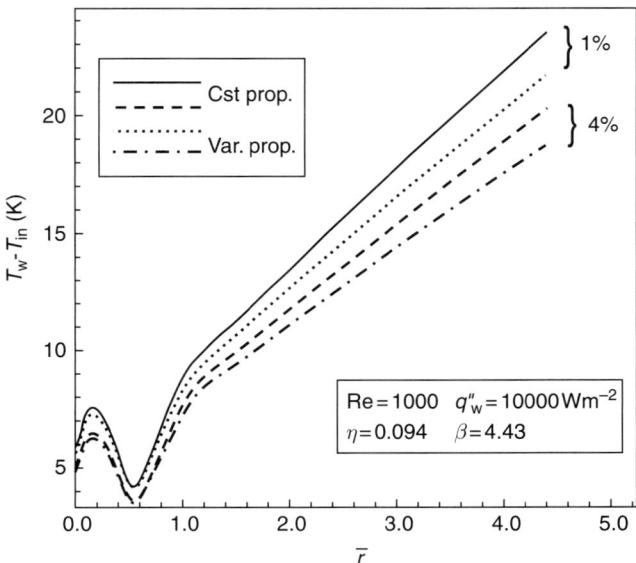

FIG. 38. Local wall temperature at different particle volumes for radial flow channel by Palm et al. [166].

Fig. 38. Also the average heat transfer coefficient increases by the use of variable properties and the wall shear stress decreases with the variable properties.

This is encouraging because both thermal and hydraulic performances are enhanced as shown in Figs. 39 and 40.

A very recent numerical study by Behzadmehr [168] brings out a vivid picture of the effect of modeling strategy on the turbulent flow simulation. They considered two modeling concepts – the mixture model and the single-phase model. In the mixture model, the fluid is considered to be a single fluid with two phases and the coupling between them is strong. But both the phases have their own velocity vectors, and within a given volume fraction there is a certain volume fraction of each phase. The governing equations are written for the mixture as

$$\nabla(\rho_m V_m) = 0 \qquad (97)$$

$$\nabla \cdot (\rho_m V_m V_m) = -\nabla p_m + \nabla \cdot (\tau - \tau_1) + \rho_m g + \nabla \cdot \left(\sum_{k=1}^{n} \phi_k \rho_k V_{dr,k} V_{dr,k}\right) \qquad (98)$$

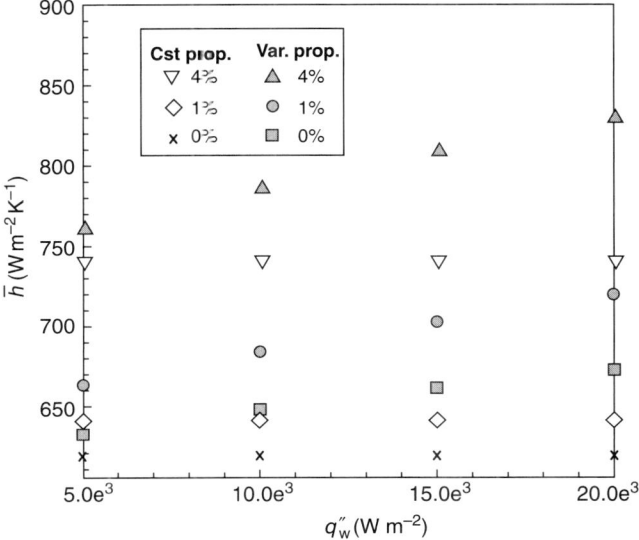

FIG. 39. Effect of variable properties on average heat transfer coefficient of a radial channel [166].

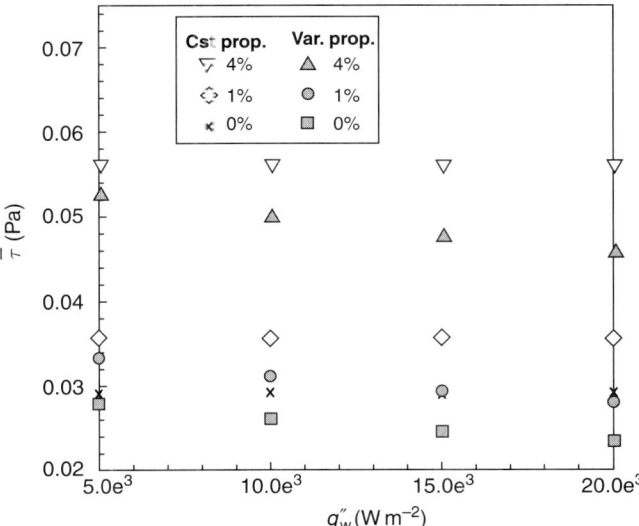

FIG. 40. Effect of variable properties on the average wall shear stress of a radial channel [166].

$$\nabla \cdot (\phi_p \rho_p V_m) = -\nabla \cdot (\phi_p \rho_p V_{dr,p}) \tag{99}$$

$$\nabla \cdot (\phi_k V_k (\rho_k h_k + p)) = \nabla \cdot (k_{eff} \nabla T - C_p \rho_m v' T') \tag{100}$$

where m stands for mixture and k for kth phase.

Here, V_{dr} is the drift velocity of phase k given by

$$V_{dr,k} = V_k - V_m \tag{101}$$

Mixture density and viscosity are given by

$$\rho_m = \sum_{k=1}^{n} \phi_k \rho_k \tag{102}$$

$$\mu_m = \sum_{k=1}^{n} \phi_k \mu_k \tag{103}$$

Shear relation is given by

$$\tau = \mu_m \nabla V \tag{104}$$

$$\tau_1 = -\sum_{k=1}^{n} \phi_k \rho_k \overline{V'_k} \overline{V''_k} \tag{105}$$

Here, V' and T' are fluctuating components of V and T in turbulent flow. The drift velocity is calculated from the relative velocity.

$$V_{pf} = V_p - V_f \tag{106}$$

where V_p is particle velocity and V_f primary phase velocity.

$$V_{dr,p} = V_{pf} - \sum_{k=1}^{n} \frac{\phi_k \rho_k}{\rho_m} V_{fk} \tag{107}$$

V_{pf} is calculated as

$$V_{pf} = \frac{\rho_p d_p^2 (\rho_p - \rho_m)}{18 \mu_f f_{drag} \rho_m} a \tag{108}$$

$$f_{\text{drag}} = 1 + 0.15\,\text{Re}\,p^{0.687}, \quad \text{Re}_p \le 1000$$
$$= 0.0183\,\text{Re}\,p, \quad \text{Re}\,p > 1000 \qquad (109)$$
$$a = g - (V_m \cdot \nabla)V_m$$

In addition to these, the equations for turbulent kinetic energy (k) and its dissipation rate (ε) are also set as

$$\nabla \cdot (\rho_m V_m k) = \nabla \cdot \left(\frac{\mu_{t,m}}{\sigma_k}\nabla k\right) + G_{k,m} - \rho_m \varepsilon \qquad (110)$$

$$\nabla \cdot (\rho_m V_m \varepsilon) = \nabla \cdot \left(\frac{\mu_{t,m}}{\sigma_\varepsilon}\nabla \varepsilon\right) + \frac{\varepsilon}{k}(C_1 G_{k,m} - C_2 \rho_m \varepsilon) \qquad (111)$$

where the turbulent viscosity is given by

$$\mu_{t,m} = \rho_m C_\mu \frac{k^2}{\varepsilon}$$

and

$$G_{k,m} = \mu_{t,m}\left(\nabla V_n + (\nabla V_m)T\right)$$
$$C_1 = 1.44,\ C_2 = 1.92,\ C_\mu = 0.09,\ \sigma_k = 1,\ \sigma_\varepsilon = 1.3$$

The boundary conditions can be set for a tube flow considering cylindrical coordinate, as at the entrance ($z = 0$)

$$Vz = V_0, \quad V = V_r = 0, \quad T = T_0, \quad \text{and} \quad I = I_0$$

where I is the turbulence intensity and I_0 is its value at the tube entrance. Under the assumption of isotropic turbulence, the turbulent kinetic energy at the entrance is given by

$$k_0 = 1.5(I_0 V_0)^2 \qquad (112)$$

At the tube outlet ($Z = L$) the diffusive flux in the axial direction should vanish. At the fluid wall interface ($r = D/Z$),

$$V_r = V_\theta = Vz = 0, \quad k = \varepsilon = 0, \quad q_w = -k_{\text{eff}}$$

With the above equations and boundary conditions, the numerical method can be carried out for the mixture model. For the single-phase model, the equations are even more simpler. It is simply the continuity, momentum, and energy equations of a single fluid with its properties replaced by the effective nanofluid properties.

Both the models were solved by the finite volume technique [169] with second-order upwind scheme used for both diffusive and convective forms. The pressure field was evolved through SIMPLE algorithm to ensure pressure–velocity coupling.

The results presented by Behzadmehr et al. [168] clearly indicated the success of the mixture model over the single-phase model in predicting the Nusselt number data produced by Xuan and Li [142] for water–Cu nanofluids.

The work also agreed with the observation of Xuan and Li [142] that the nanoparticles do not have significant effect on the frictional behavior of the fluid.

Another work which questioned the role of uncertainties in physical properties on convective heat transfer in nanofluids was carried out by Mausour et al. [170]. For assessing whether the performance of a nanofluid while replacing a base fluid during a cooling exercise, one needs to examine the variation in pumping power and mass flow rate for a given heat transfer or variation of bulk fluid temperature or pumping power for a given heat transfer rate and mass flow rate. In engineering terms, this is "rating" of the cooling arrangement or more simply it is the assessment of hydraulic penalty (power consumption due to pressure drops) for a given tube geometry and heat flux.

Mansour et al. [170] showed that the nature of variation of these parameters (power ratio or mass flow rate) depends critically on which models of effective properties are used. As it is not yet very certain which models for viscosity, specific heat, and thermal conductivity are actually applicable to nanofluids, they took two sets of equations naming them GdS (Gosselin and da Silva) and BMGN (Ben Mansour and Galanis Nicolas). In GdS model, they used Pak and Cho [137] model for specific heat

$$C_p = (1 - \phi_p)(C_p)_f + \phi_p(C_p)_p \qquad (113)$$

where f indicates base fluid and p particle. They used Brinkman [171] model of viscosity for GdS model.

$$\mu_{\text{eff}} = \mu_f \frac{1}{(1 - \phi_p)^{2.5}} \qquad (114)$$

Finally, for GdS model, they use the Hamilton–Crosser model for spherical particles as

$$\frac{k_{\text{eff}}}{k_0} = \frac{k_p - 2k_f - 2\phi_p(k_f - k_p)}{k_p - 2k_f - \phi_p(k_f - k_p)} \qquad (115)$$

For the BMGN model, they used the Wang et al. [30] model for viscosity, Xuan and Roetzel [152] model for specific heat, and Yu and Choi [93] for thermal conductivity as

$$\frac{\mu_{\text{eff}}}{\mu_{\text{f}}} = 123\phi_p^2 + 7.3\phi_p + 1 \tag{116}$$

$$(\rho C_p)_{\text{eff}} = (1-\phi_p)(\rho C_p)_f + \phi_p(\rho C_p)_p \tag{117}$$

$$\frac{k_{\text{eff}}}{k_f} = \frac{k_p + 2k_f + 2(k_p - k_f)(1+\beta)^3 \phi_p}{k_p + 2k_f - 2(k_p - k_f)(1+\beta)^3 \phi_p} \tag{118}$$

Their results showed that for a fixed mass flow rate, even though the difference between wall and bulk temperatures decreases for both the models (Fig. 41), for the pumping power ratio (Fig. 42), two models show completely different trends – one showing increase in pumping power with volume fraction, while the other showing decrease.

The above results clearly indicate that before concluding the thermal and hydraulic performances of convection with nanofluids, we must have accurate model for property evaluation.

Another important work on this was by Gosselin and de Silva [172]. They tried to optimize the particle loading for laminar and turbulent forced and natural convections in nanofluids. They defined the optimum as the highest value of relative heat transfer Ω_{FC} (heat transfer in nanofluid divided by heat

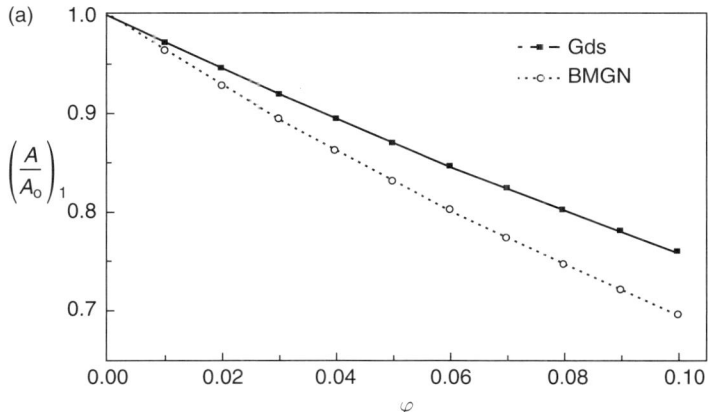

FIG. 41. Effect of particle loading on difference between wall and bulk temperatures [170].

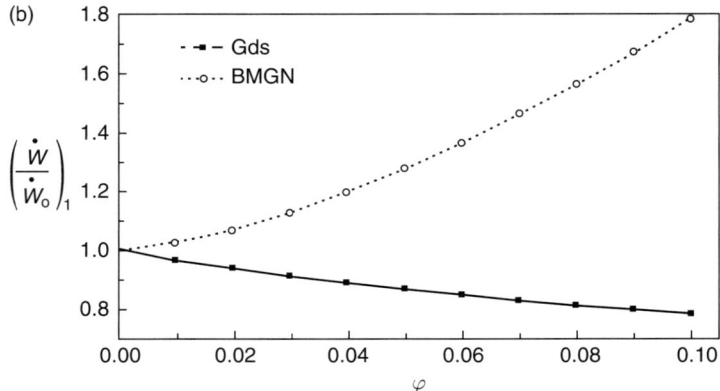

FIG. 42. Effect of particle loading on pumping power for fixed heat rate and mass flow rate [170].

transfer in base fluid) for a fixed pumping power. For laminar and turbulent forced flow, it is given by

$$\Omega_{FC,laminar} = \frac{\tilde{k}_{eff}^{2/3} \tilde{\rho}_{eff}^{2/5} \tilde{C}_{p,eff}^{1/3}}{\tilde{\mu}_{eff}^{4/15}} \tag{119}$$

$$\Omega_{FC,turbulent} = \frac{\tilde{k}_{eff}^{2/3} \tilde{\rho}_{eff}^{4/7} \tilde{C}_{p,eff}^{1/3}}{\tilde{\mu}_{eff}^{11/21}} \tag{120}$$

where

$$\tilde{k}_{eff} = \frac{k_{eff}}{k_f}$$

$$\tilde{\rho}_{eff} = \frac{\rho_{eff}}{\rho_f}$$

$$\tilde{\mu}_{eff} = \frac{\mu_{eff}}{\mu_f}$$

For laminar natural convection, it is

$$\Omega_{NC,laminar} = \frac{\tilde{\beta}_{eff}^{1/4} \tilde{\delta}_{eff}^{1/2} \tilde{C}_{p,eff}^{1/4} \tilde{k}_{eff}^{3/4}}{\tilde{\mu}_{eff}^{1/4}} \tag{121}$$

This work clearly demonstrates that an optimization in particle loading is possible. At the same time, it commented that the entire exercise is limited by the accuracy of the correlations used in the optimization process which is similar conclusion as that of Mansour et al. [170].

Thus, the numerical works as discussed above considered a whole range of theoretical concepts for simulation. This includes the two-phase and single-phase models, dispersion model, and Eulerian and Lagrangian approaches. One of the major issues in these simulations is found to be the evaluation of thermophysical properties in general and viscosity and thermal conductivity in particular for the nanofluids. Use of classical models such as Einstein or Brinkman model for viscosity and Maxwell and Hamilton–Crosser model for thermal conductivity is questionable for nanofluids. On the other hand, the available experimental data on nanofluids is too little to converge into one model for them.

E. Natural Convection in Nanofluids

The natural convection of suspensions finds important application in chemical, pharmaceutical, food and beverage, and refrigeration industry as well as in solar collectors. The experimental papers are scarce in the field of natural convection with nanofluids as compared to the forced convection. The initial studies in natural convection of nanofluids were carried out by Putra et al. [167]. They investigated on natural convective heat transfer of aqueous CuO and Al_2O_3 nanofluids inside a horizontal cylinder heated from one end and cooled from the other. The dependence of parameters such as particle concentration, particle material, and geometry of the containing cylinder was also investigated. Unexpectedly and different from the results of thermal conduction and forced convection, the presence of nanoparticles has been found to deteriorate the natural convective heat transfer systematically. The experiments at Rayleigh number from 10^6 to 10^9 showed a significant deterioration in natural convective heat transfer. For instance, for 4% weight concentration of CuO and alumina nanofluid, approximate decrease of 250% and 150% in Nusselt number was observed at a given Rayleigh number of 5×10^7, respectively. Figure 43 shows the above variation.

The deterioration increased with particle concentration and was more prominent for CuO nanofluids. The aspect ratio of the cylinder was also observed to influence heat transfer greatly. The decrease in heat transfer has been observed to be more severe for aspect ratio of 1.0 than that of 0.5. The reasons for the above deterioration in heat transfer were stated to be due to particle–fluid slip and sedimentation.

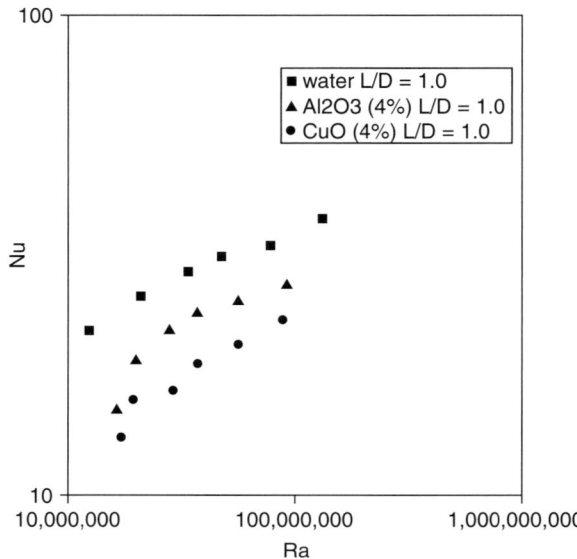

FIG. 43. Nu against Ra for various nanofluids by Putra *et al.* [167].

Following this, another experimental study on natural convection was carried out by Wen and Ding [173]. They suspected that the deterioration in the studies of Putra *et al.* [167] were due to use of non stabilized nanofluid and they conducted a through stabilization procedure. Titanium dioxide nanoparticles were dispersed in distilled water through electrostatic stabilization mechanisms and with the aid of a high shear mixing homogenizer. Nanofluids formulated in such a way were found very stable and were used in the investigation of their heat transfer behavior under the natural convection conditions. A meticulous experimental investigation was carried out; however, the experimental did agreed with the observations of Putra *et al.* [167]. A typical experimental result depicting the deterioration is shown in Fig. 44.

Both transient and steady heat transfer coefficients were obtained for different concentrations of nanofluids under natural convective conditions. It was also observed that the deterioration increases with nanoparticle concentrations. Possible reasons for deterioration in heat transfer coefficient were attributed to due to facts such as convection induced by concentration difference, particle–surface and particle–particle interactions, and modifications of the dispersion properties.

With the serious lack in experimental results in natural convection, investigators tried to numerically predict the nanofluid behavior under free

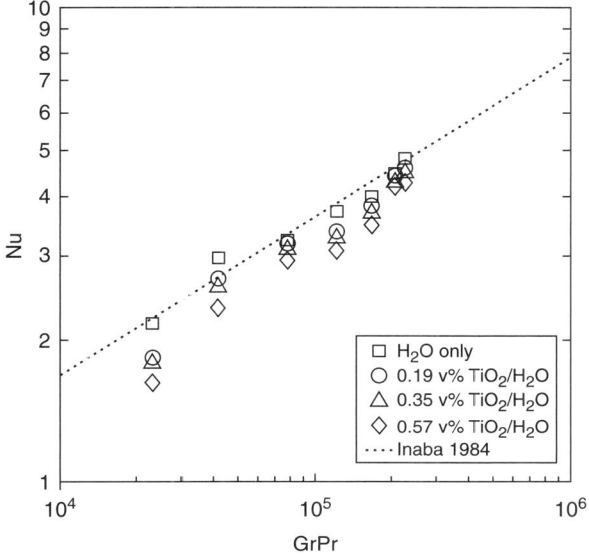

FIG. 44. Deterioration in natural convection by Wen and Ding [173].

convection. It may be observed that many of the numerical predictions were not matching with the experimental results. This may be due to the use of major assumptions in the analysis such as homogenous mixer assumption, no-slip between particle and fluid, and varying thermophysical properties. But the efforts and methodology used in modeling natural convection is quiet appreciable and is reported here in detail.

Khanafer *et al.* [174] numerically investigated the heat transfer behavior of nanofluids in a two-dimensional horizontal enclosure. The nanofluid in the enclosure was assumed to be in single phase, that is, both the fluid and particles are in thermal equilibrium and flow at the same velocity. The effect of suspended nanoparticles on the buoyancy-driven heat transfer process was analyzed. They used the well-known stream function–vorticity formulation. The random motion of nanoparticles was considered through a dispersion model similar to thermal dispersion models for flows through porous media. The energy equation in the analysis takes the form as given in Eq. (122).

$$\frac{\partial T}{\partial t} + u\frac{\partial T}{\partial x} + v\frac{\partial T}{\partial y} = \frac{\partial}{\partial x}\left[\left(\alpha_{\mathrm{nf}} + \frac{k_{\mathrm{d}}}{(\rho C_{\mathrm{p}})_{\mathrm{nf}}}\right)\frac{\partial T}{\partial x}\right] + \frac{\partial}{\partial y}\left[\left(\alpha_{\mathrm{nf}} + \frac{k_{\mathrm{d}}}{(\rho C_{\mathrm{p}})_{\mathrm{nf}}}\right)\frac{\partial T}{\partial y}\right]$$
(122)

It can be seen that along with the nanofluid conductivity a dispersive conductivity k_d is also assumed. It is modeled following the theory of porous medium and is given as

$$k_d = C(\rho C_p)_{nf}|\overline{V}|\phi d_p \tag{123}$$

where $|\overline{V}| = \sqrt{u^2 + v^2}$, C is a constant, and u and v are axial and radial velocities.

They used the Brinkman [171] model for viscosity and Wasp [14] model for thermal conductivity of nanofluids. They used the finite difference technique with alternate direction implicit (ADI) algorithm with power law scheme to solve the transient equations and validated them by comparing with the solutions of FIDAP software as well as experimental value of pure fluids. Subsequently, they carried out studies on natural convection in the differentially heated cavity with nanofluids. Considerable difference in velocity and temperature was obtained between pure fluids and nanofluids as shown in Fig. 45.

The simulations showed that suspended nanoparticles substantially increased heat transfer at any given Grashof number. Such enhancement increased with particle concentration, which was thought to be the increased energy exchange from enhanced irregular and random movements of particles. They also presented a correlation for average Nusselt number in the form:

$$\overline{Nu} = 0.5163(0.4436 + \phi_p^{1.0809})Gr^{0.3123} \tag{124}$$

where Gr is the Grashof number and ϕ_p the particle volume fraction. A similar work was carried out by Jou and Tzeng [175] inside a differentially heated cavity. They also used the stream function–vorticity formulation in a way identical to the previous study by Khanafer *et al.* [174]. They presented, in addition to the Grashof number effects, the effect of the cavity aspect ratio (width/height) on the thermal behavior. Figure 46 shows the effect of aspect ratio on the isotherms for natural convection in nanofluids with 20% particle concentration. However, the use of 20% volume fraction is questionable in practice, as it is extremely difficult to make a stable nanofluid in that concentration. Also at such volume fractions, the Newtonian behavior of the fluid is doubtful.

Kim *et al.* [176] proposed an analytical investigation to describe the natural convective heat transfer of nanofluids by considering the effect of the ratio of thermal conductivity of nanoparticles to that of the base fluid,

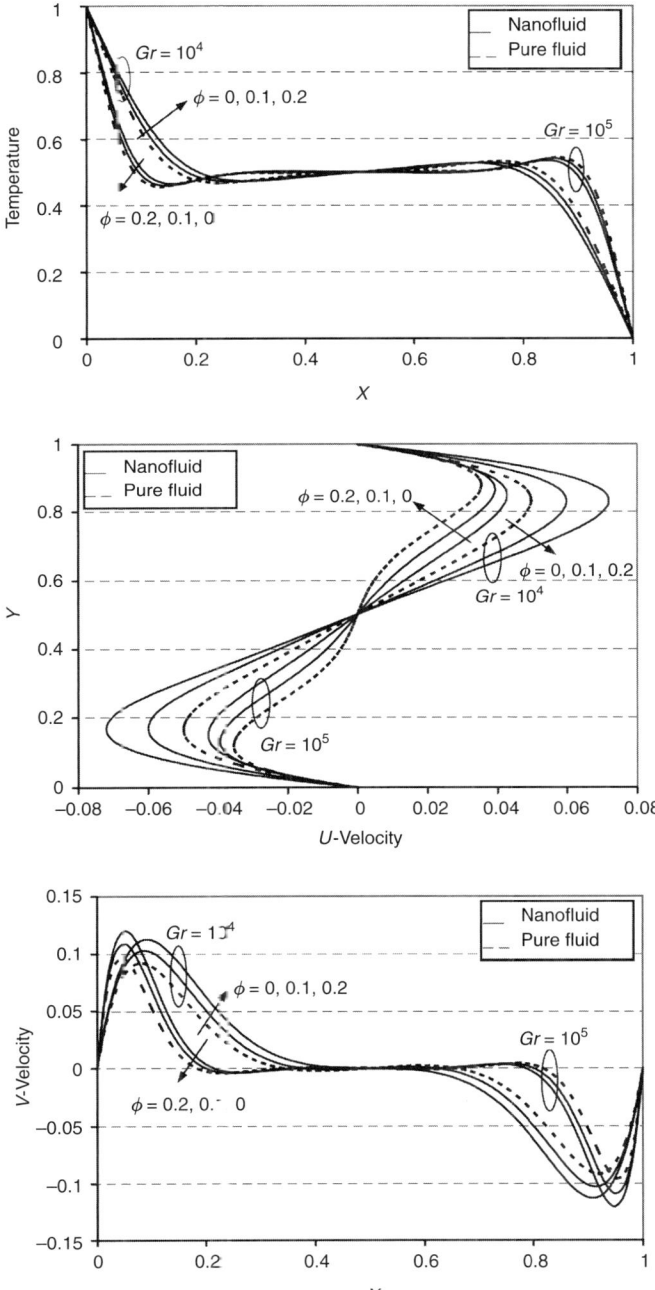

FIG. 45. Comparison of temperature and velocity profiles between nanofluid and pure fluid [174].

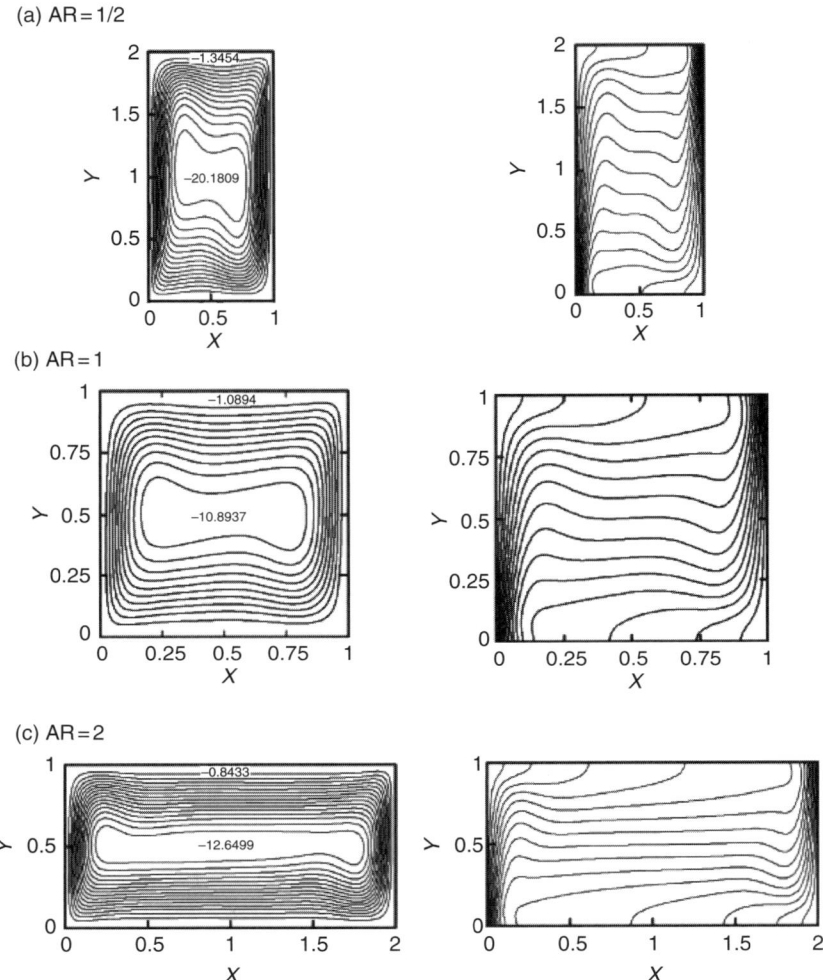

FIG. 46. Stream line and isotherm at various aspect ratios by Jou and Tzeng [175] (Gr = 10^5, Pr = 6.2, vol fraction 20%).

the shape factor of the particles, the volume fraction of nanoparticles, the ratio of density of nanoparticles to that of the base fluid, and the ratio of heat capacity based on the volume of nanoparticles to that of the base fluid. The results showed that the heat transfer coefficients of nanofluid increased with increasing particle volume fraction.

Kim et al. [177] further advanced their study on the stability of nanofluids with binary fluids $H_2O/LiBr$ and NH_3/H_2O which are used in the absorption refrigeration systems. They considered linear temperature gradient giving linear concentration gradient in the basic state as

$$C_c = C_i\{1 - \beta_T \Delta(d - z)\} \tag{125}$$

where C is the concentration, β_T is the temperature gradient, d is the fluid layer thickness, and z is the vertical coordinate from the bottom plate. They considered both Soret effect (particle diffusion under temperature gradient) and Dufour effect (heat transfer induced by concentration gradient). The heat and mass flows with these effects are given by

$$-J_h = k\nabla T + \rho c_p \alpha_{Df} \nabla C \tag{126}$$

$$-J_m = D\nabla C + D_{sr}\nabla T \tag{127}$$

Here, α_{Df} and D_{sr} are Dufour and Soret coefficients, respectively. The governing equations were developed as

$$\nabla \vec{U} = 0 \tag{128}$$

$$\rho_R \frac{D}{Dt}\vec{U} = -\nabla P + \mu \nabla^2 \vec{U} + \rho \vec{g} \tag{129}$$

$$\frac{DT}{Dt} = \alpha \nabla^2 T + \alpha_{Df} \nabla^2 C \tag{130}$$

$$\frac{DC}{Dt} = D\nabla^2 C + D_{sr}\nabla^2 T \tag{131}$$

$$\rho = \rho_R[1 - \beta_T(T - T_R) + \beta_S(C - C_R)] \tag{132}$$

The above equations are for continuity, momentum, energy, and concentration, and D/Dt is the total differential

$$\frac{D}{Dt} = \frac{\partial}{\partial t} + \vec{U}.\nabla \tag{133}$$

From these equations, they derived the liner stability equation.

$$(D^2 - a^2)^3 w^* = \overline{Ra}\, a^2 w^* \tag{134}$$

where

$$\overline{Ra} = Ra(1 + F_s + F_{sr} + F_{Df})(1 - K)^{-1} \tag{135}$$

$$Ra = \frac{g\beta_T \Delta T d^3}{\alpha \gamma}$$

$$F_s = \frac{\beta_s C_i \alpha}{D}$$

$$F_{sr} = \frac{\beta_s D_{sr}}{\beta_T D}$$

$$F_{Df} = \frac{\beta_T C_i \alpha_{Df}}{D}$$

$$K = \frac{\alpha_{Df} D_{sr}}{\alpha D}$$

With the above parameters, they calculated the stability parameters for different systems of nanofluids (copper and silver). The separation factor, which is a prime stability parameter, is given by the following equation for no Dufour effect

$$\psi = \frac{(1 - \phi_w) + \phi_w \delta_4}{(1 - \phi_w) + \phi_w \delta_3} \psi_{bf} \tag{136}$$

where ψ_{bf} is the separation factor for the base fluid, ϕ_w is the weight fraction of particles, and

$$\delta_3 = \frac{D \text{ nano particles}}{D_s}$$

$$\delta_4 = \frac{D_{sr},\ \text{nano particles}}{D_{sr},\ \text{solute in binary fluid}}$$

They concluded that for Dufour and Soret effect makes nano-fluid unstable and for heat transfer, Soret effect is more important. They also concluded that denser initial concentration makes the nanofluid more unstable.

Recently, Tiwari and Das [178] conducted a numerical study on mixed convection in two-sided lid-driven differentially heated square cavity filled with nanofluid. The nanofluid in the enclosure was assumed to be Newtonian, incompressible, and laminar, and also had a uniform shape and size. Moreover, it was assumed that both the fluid phase and nanoparticles were in thermal equilibrium state and they flowed at the same velocity. The thermophysical properties of the nanofluid were assumed to be constant except for the density variation in the buoyancy force, which were based on the Boussinesq approximation. In the study, a time marching incompressible flow solver has been applied for simulating the flow features of nanofluids for a range of solid volume fraction (χ) values and Richardson number (Ri). Three cases were analyzed. In case I, the left wall (cold) is moving up while the right wall (hot) is moving down. In case II, the left wall is moving down while the right wall is moving upward, and in case III, both the walls are moving upward. A typical result for case I, in which the left wall (cold) is moving up while right wall (hot) is moving down and with $Ri = 10$, is shown in Fig. 47.

It was observed that the nanoparticles when immersed in a fluid are capable of increasing the heat transfer capacity of base fluid. As solid volume fraction increases, the effect is more pronounced. Nanoparticles were able to change the flow pattern of a fluid from natural convection to forced convection regime. It was also seen that when both the vertical walls move upward in the same direction, the heat transfer was reduced compared to the other two cases.

A theoretical note on the heat transfer behavior of Newtonian nanofluid in laminar natural convection regime was recently given by Polidori et al. [179]. Here, a laminar external boundary layer was investigated from the integral formalism approach. The major point which was tried to be emphasized in the paper was the fact that natural convective heat transfer was not solely characterized by the nanofluid effective thermal conductivity, while the sensitivity to the viscosity model used also played a key role in the heat transfer behavior. It was also observed that the use of the Brinkman model for the dynamic viscosity yields a systematic and significant heat transfer enhancement, regardless of the particle concentration. On the other hand, the use of the experimental correlation for the viscosity leads only to a weak enhancement (less than 1%) with a trend to a deterioration phenomenon with increasing particle concentration in the considered range. Further semianalytical formulas of heat transfer parameters have been proposed

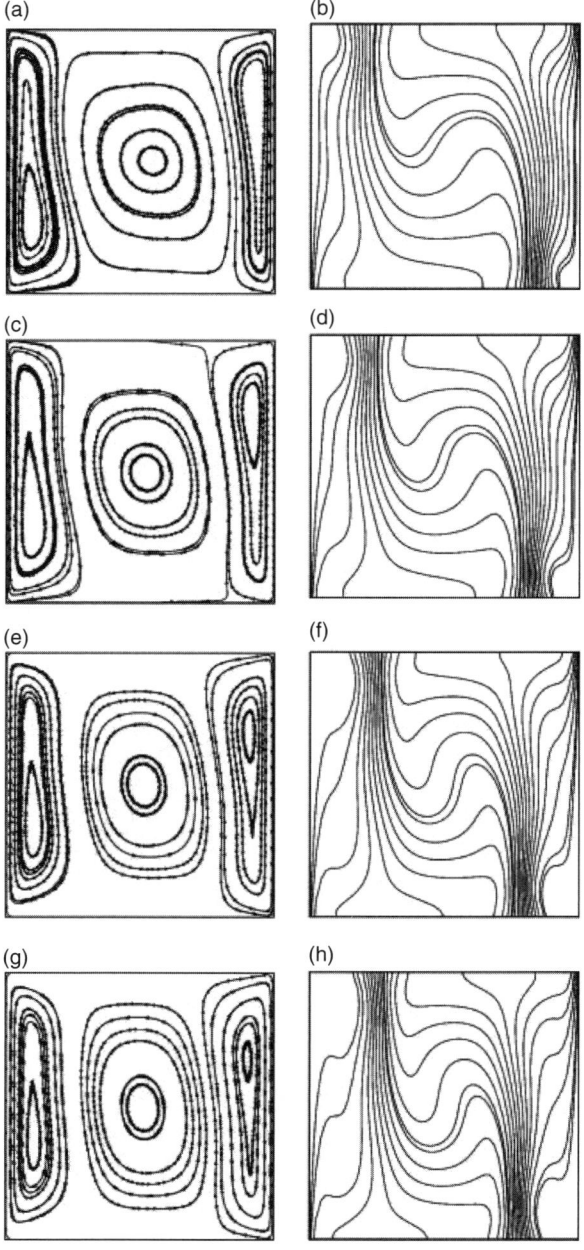

FIG. 47. Streamlines and isotherms, $Ri = 10$: (a) $\chi = 0\%$; (b) $\chi = 0\%$; (c) $\chi = 8\%$; (d) $\chi = 8\%$; (e) $\chi = 16\%$; (f) $\chi = 16\%$; (g) $\chi = 20\%$; and (h) $\chi = 20\%$ [178].

for both the uniform wall temperature (UWT) and the uniform heat flux (UHF) surface thermal conditions. The major disadvantage of the above work was the fact that in their analysis constant property values were assumed which are highly questionable.

In yet another work, Akbarinia [180] investigated on laminar mixed convection of a nanofluid consisting of water and Al_2O_3 under buoyancy and heat transfer in a curved tube. Simultaneous effects of the buoyancy force, centrifugal force, and nanoparticles concentration on the fluid flow and heat transfer along the pipe were investigated considering three-dimensional elliptic governing equations. It was observed that for a given heat flux, increasing centrifugal force augments Nusselt number but reduces the skin friction factor and for a given heat flux at a given flow rate, skin friction factor is increased and Nusselt number is reduced by augmenting nanoparticles concentration.

It may be noted here that all the numerical and analytical predictions gave enhanced heat transfer coefficient contrary to the experimental results. Thus more extensive experimental investigations are required to conclude with the real nature of heat transfer.

The path seems to be long and the journey has just begun. There are number of questions remains to be answered as far as fluid mechanics and heat transfer of nanofluids are concerned. How does the viscosity change in metallic nano-fluids? Does the micro- or nanoconvection interfere with the imposed bulk convection? What role does particle migration mechanisms such as thermophoresis or diffusionphoresis play? Does the transition from laminar to turbulent flow change? What is the effect of buoyancy on the particles? The stream of questions seems to have just started emerging.

XI. Boiling in Nanofluids

Boiling is one of the most efficient modes of heat transfer and hence is used in a wide range of applications. Even though many researchers worldwide have studied the basic mechanism of boiling, it still remains too complex physical mechanism to be completely understood even for a common fluid like water. It is known to depend mainly on surface heat flux, heater surface, and heater geometry. Also, it is known that the inclusion of particles in a liquid alters the boiling characteristics. Yang and Maa [181] performed pool boiling experiments with alumina–water solid particle suspensions. They used Al_2O_3 particles of sizes 50 nm, 300 nm, and 1 μm. They found that pool boiling performance is greatly improved for low particle concentrations of 0.1–0.5% in nucleate pool boiling regime. However, micron-sized particle suspensions are known to cause problems of erosion and clogging.

The thermal conductivity enhancement in nanofluids gave a hope that these fluids can also be advantageous in boiling since conduction plays a major role in the boiling process. However, quite often the reality in science and technology goes against intuition as can be observed in the section below.

A. Pool Boiling Heat Transfer at Higher Solid Particle Concentrations

The pioneering work in pool boiling of nanofluids was by Das et al. [139] who investigated nucleate pool boiling characteristics of Al_2O_3–H_2O nanofluids on a cylindrical cartridge heater. The thrust of the experiments was to compare the pool boiling parameters with that of pure water and thus bring out the applications and limitations of nanofluids under the condition of phase change.

In their experiments, stainless steel heaters of 20 mm diameter and 420 V, 2.5 kW rating was used. The heater surface was machine drawn. They conducted experiments with high solid particle concentrations of 4–16% by weight. In this work, neither the nanofluids were electrostatically stabilized nor was surfactant used to stabilize the nanofluid. The higher the concentration the more was the sedimentation and hence the boiling performance worsened. This is shown in Fig. 48. The nanoparticles were found to sediment on the heater, thus making it smoother and deteriorating the boiling performance (Fig. 48).

This brings out the probable cause for the deterioration in boiling characteristics. Due to the fact that the size of the nanoparticles (20–50 nm) is 1–2 orders of magnitude smaller than the roughness (0.2–1.2 μm) of the heating surface, the particles sit on the relatively uneven surface during boiling. These trapped particles change the surface characteristics, making it smoother. This causes the degradation of the boiling characteristics. For higher particle concentration, the particle virtually forms a layer on the heating surface, hindering the heat transfer. Thus, the small size of the particles causes the surface skirting which overshadows the thermal conductivity enhancement of the nanofluids.

Later on, the same authors Das et al. [182] showed (Fig. 49) that pool boiling of nanofluids on narrow horizontal tubes (4 and 6.5 mm diameter) is qualitatively different from the large diameter tubes due to difference in bubble sliding mechanism. It was found that at this range of narrow tubes the deterioration in performance in boiling of nano-fluids is less compared to large industrial tubes, which makes it less susceptible to local overheating in

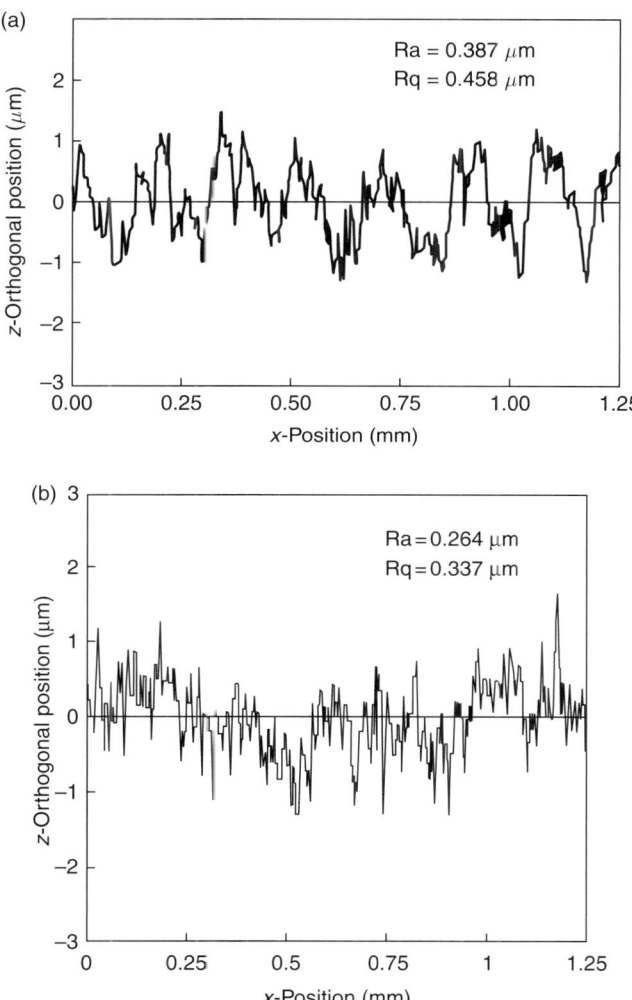

FIG. 48. (a) Surface characteristics before boiling [139]. (b) surface characteristics after boiling [139].

convective applications. For boiling on tubes of 4 and 6.5 mm diameter, there seems to be less importance of sliding mechanism for larger bubbles, which are comparable to the size of bubbles of boiling on 20-mm tube.

This is because of the relatively small size of the tube, which produces a large curvature of the surface, which does not allow the sliding of larger bubbles but induces direct departure. However, a large number of smaller bubbles are produced in a sustainable way here and they slide but to a relatively smaller distance.

Figure 50 clearly indicates that, in general, the boiling performance of the base fluid deteriorates with the addition of nanoparticles pushing the boiling curves to the right, which means that the nanofluid can cause harm to cooled surface if boiling occurs because it will give a higher surface temperature compared to water at the same heat flux similar to that observed earlier [139].

It has been observed that the shift of the curve to the right is not proportional to the particle concentration and it is strongly dependent on the tube diameter even for the similar values of surface roughness. For narrower heaters (4 and 6.5 mm), the shift of the curve is considerable and is almost of the same order over the entire range of heat fluxes. For 20-mm tube from 1% to 4% concentration, a regular shift of the curve was observed at lower heat fluxes but at the upper part of the curves the difference between wall

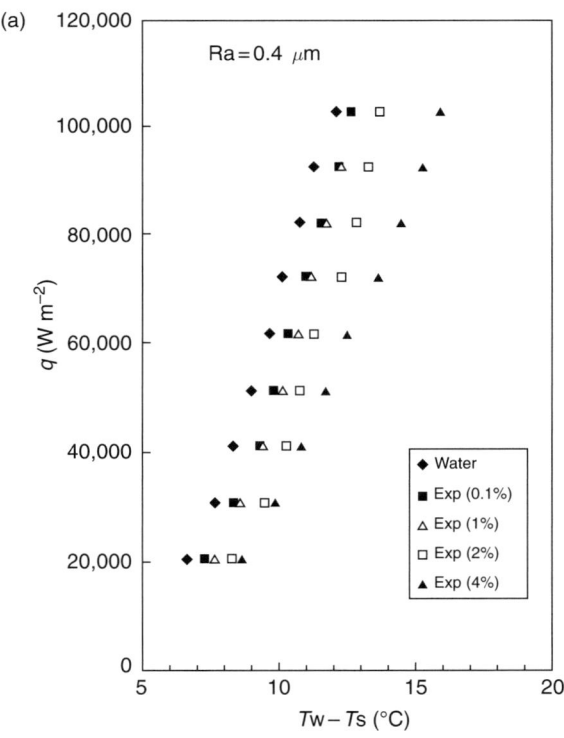

FIG. 49. (Continued)

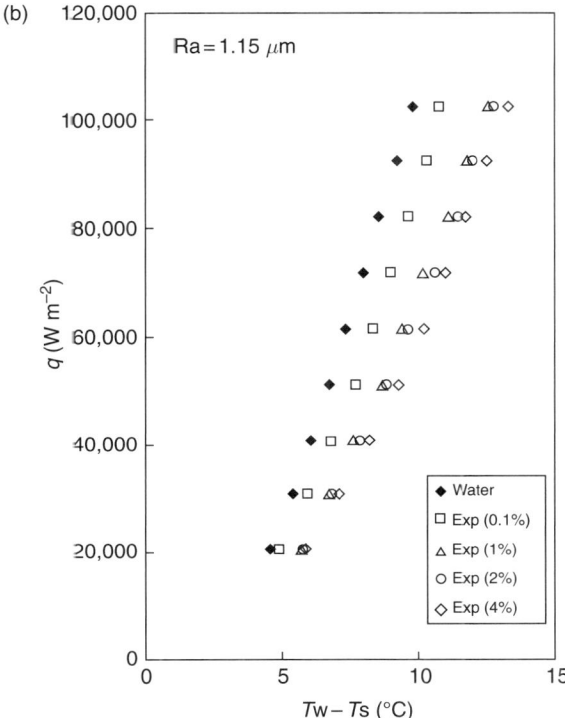

FIG. 49. (a) Results of Das et al. [139]. (b) Results of Das et al. [139].

superheats for various particle concentrations was found to increase with increasing heat flux.

To further understand the effect of nanoparticles on heat transfer, dimensionless Nu–Re_b plot was presented in Fig. 51. This figure indicates that for each particle concentration the Nu–Re_b characteristics are different and shifted downward. This is a general observation for all the tubes which indicates that the change in boiling characteristics of nanofluids can be explained neither in terms of property change nor in terms of changes in Nu and Re_b due to change in characteristic length (diameter). The change in Nu–Re_b correlations is more drastic at higher Re_b for large-diameter tubes than for narrow tubes. This indicates that the danger of local overheating is smaller on narrow tubes compared to larger tubes for high heat flux applications when boiling point is reached for a nanofluid.

Bang and Chang [183] studied pool boiling characteristics of Al_2O_3–H_2O nanofluids at higher heat fluxes and smoother heaters. Their experiments were also with 4–16 wt% nanofluids. They used much smoother heater

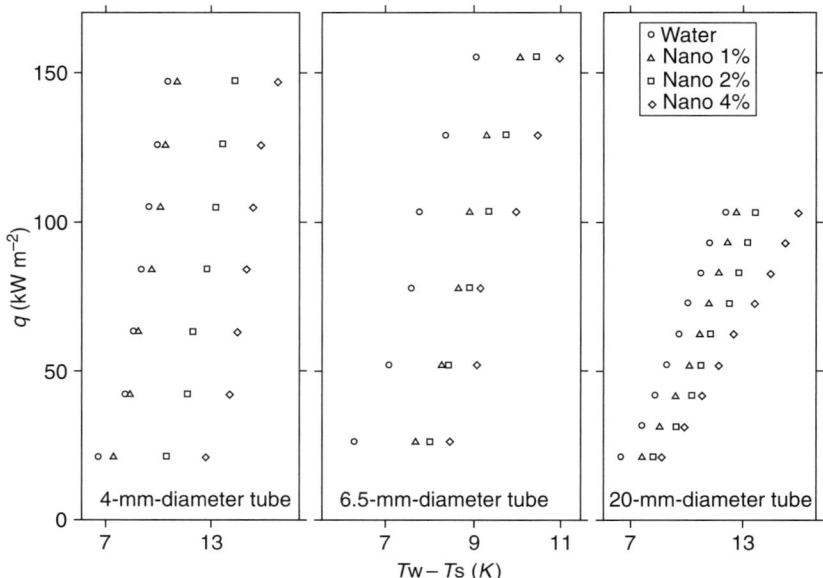

FIG. 50. Results of Das *et al.* [182].

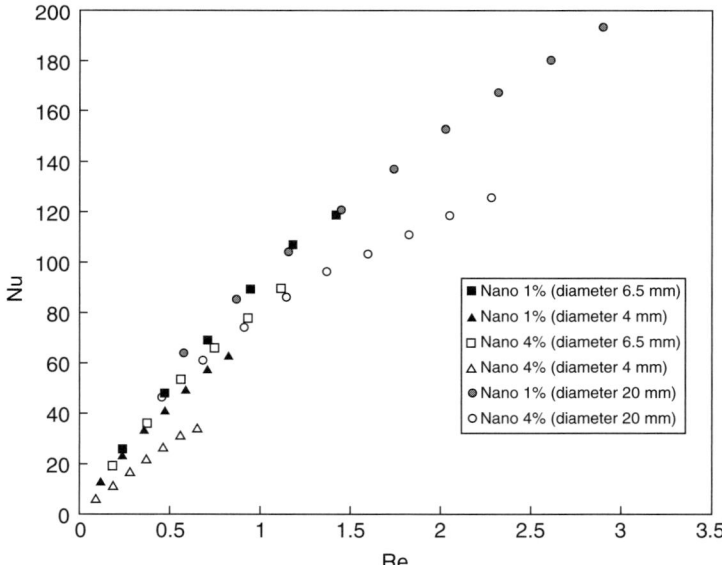

FIG. 51. Dimensionless boiling characteristics of nanofluids boiling on narrow tubes [182].

compared to Das *et al.* [139] having surface roughness of ~37 nm. They had some important observations regarding the boiling characteristics of nanofluids. First, they also observed deterioration of boiling with nanofluids concentration in nanofluids similar to Das *et al.* [139] but the rate of heat transfer was somewhat different which they attributed to the difference in geometrical features of the heaters in the two studies. They could also identify a clear natural convection regime followed by nucleate boiling as shown in Fig. 52. They further observed that the experimental data does not conform to the Rohsenow correlation, just by changing the properties of the fluid with effective nanofluid properties.

They tried different variations of the same correlation like using Rohsenow correlation with changing only the effective conductivity or changing the constant C_{sf} of the Rohsenow correlation. It was found that rather than changing the properties of the fluid, the modification of the surface–fluid combination factor, C_{sf}, gives closer approximation to the experimental boiling data of nanofluids. This definitively indicates that the modification of surface characteristics during the boiling of nanofluids might hold the key in explaining the deterioration of boiling of nanofluid. However, the explanation of Bang Chang [183] regarding the surface modification was quite different.

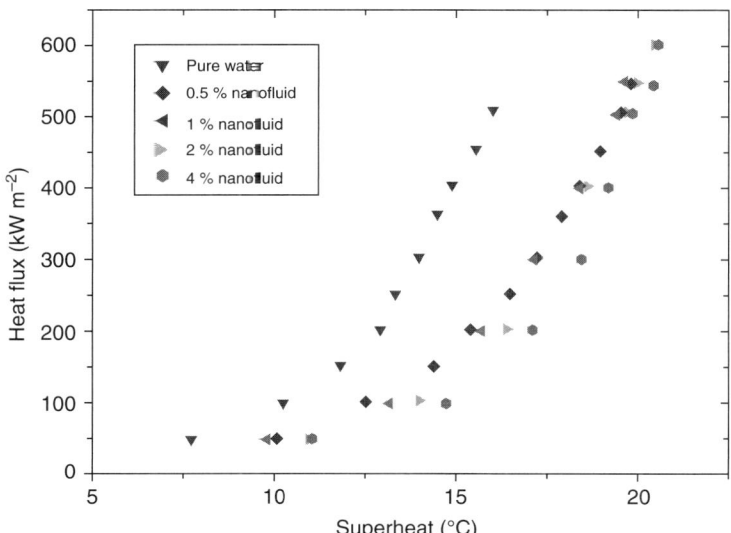

FIG. 52. Boiling curve for nanofluids from Bang and Chang [183].

Contrary to the observation of Das *et al.* [139], they found that the roughness of the clear heater was increased after boiling of 0.5% alumina nanofluid and even more increased in 4% alumina nanofluid. They found locally smoothened heater also. They also explained the result that due to the closeness of roughness and particle size here, the fouling effect (particles forming a layer on the surface) is overriding which reduces heat transfer due to poor conductivity of Al_2O_3. This reduces the heat transfer even though the surface roughness is increased. They commented that if the surface roughness is much higher than the particle size, the plugging effect of nucleation sites are to be expected as observed by Das *et al.* If the roughness is smaller than the particle size, the increase of roughness and formation of layer is expected as observed by Bang and Chang [183]. Thus we find that here again the nanoparticles were found to foul the surface, causing a decrease in pool boiling heat transfer coefficient. Although they found that the fouling layer gave higher wall surface roughness readings, when it was locally cleaned, the heater was found to be smoother than before the experiment. Thus, in general, a deterioration in boiling was observed for nanofluids with higher particle concentration. However, Das *et al.* [139] indicated that even the deterioration can have practical application in having "engineered fluids" which will inhibit boiling or will boil at a preassigned surface temperature which may be important for heat treatment or material processing. This has been proved to be correct when Tsai *et al.* [184] used nanofluids in heat pipes to delay the boiling limit.

B. Pool Boiling Heat Transfer at Lower Solid Particle Concentrations

The results of Wen and Ding [185] gave a completely different picture of boiling of nanofluids. They observed an enhancement of boiling in the presence of nanoparticles. The particles used by them were same as those used by Das *et al.* [139] acquired from the same company (nanophase technologies) with particle sizes of 10–50 nm. They stabilized the suspension by adjusting the pH value near 7, which is away from the isoelectrical point (IEP) of alumina (about 9.1). They also used a high-speed homogenizer (~24,000 rpm) for breaking the agglomerates of Al_2O_3 powder. Even after these processes, they found considerable agglomeration giving an average particle size of 167.54 nm but the nanofluid was stable. They used 2.4-kW ring heater below stainless steel boiling surface.

Their results were quite different from the earlier studies. While their pure water results matched with the traditional Rohsenow correlation, the heat transfer with nanofluids showed an enhancement in heat flux at the same wall superheat and this enhancement increased with particle volume fraction. They observed an increase in heat transfer coefficient as high as 40%. Also, the enhancement was with just 1.25 wt% of particles, which is about 0.3% by volume. They found that this increase is much more than the measured value of thermal conductivity enhancement and hence the boiling enhancement cannot be explained by conductivity enhancement alone. They also observed that the enhancement, apart from being strongly dependent on the particle concentration, is also dependent on heat flux giving higher values with increasing heat flux.

The explanations of the above results were not conclusive because the authors themselves only indicated several possibilities. They first indicated the possibility of agglomeration remaining in the fluids used by Das et al. [139] and Bang and Chang [183]. However, it must be indicated that the particle concentration of Wen and Chang [185] was much less than (an order of magnitude lower) Das et al. [139] and Bang and Chang [183]. While Das et al. [139] used 1–4% particles by volume, Wen and Ding [185] used 0.32–1.25% by weight (which is about 0.08–0.3% by volume). In fact looking at the above results, it may be said that although the results are different with respect to enhancement of boiling (one giving positive, another negative enhancement), they need not negate each other and both results may be true in the respective ranges of particle concentration.

These two ranges may be dominated by different phenomena giving different heat transfer characteristics. Wen and Ding [185] also indicated other reasons such as surface characteristics other than roughness such as surface wettability, effect of dispersants and surfactants, measurement techniques, and characteristic size of the system as the probable explanations for the discrepancy in the results of pool boiling of nanofluids. There are a few more works at lower particle concentrations. You et al. [186] performed experiments with silica–water and alumina–water nanofluids of very small solid particle concentrations (0.0001–0.005% by weight) on a 10–mm^2 heater in subatmospheric conditions. They found no significant change in nucleate pool boiling.

Vasallo et al. [187] conducted experiments with silica–water nanofluids (2% by weight) of different particle sizes, ranging from 15 nm to 3000 nm on a NiCr wire heater, and found no significant change in the boiling performance at low and medium heat fluxes. But at heat fluxes near to critical heat flux (CHF) of water, they observed that there is boiling deterioration for the 50-nm nanofluid. Witharana [188] carried out experiments using gold–water nanofluids of very low solid particle concentrations (0.001% by weight) on plate heater. An enhancement of 11–21% in heat transfer coefficient was found. With increasing particle concentration, the percentage enhancement in heat transfer coefficient also increased. Thus, it is clear that although pool boiling heat transfer using nanofluids has been a subject of several investigations in the past few years, incongruous results have been reported in literature

TABLE III

	Heat transfer coefficient	Heater used	Nanofluid
Yang and Maa [9]	↑↑	3.2-mm horizontal tube heater	Micron sized Al_2O_3 particles suspended in water (0.4–2 wt%)
Das et al. [1]	↓↓	Cylindrical cartridge heater of 20-mm diameter	Al_2O_3–water nanofluid (4–16 wt%)
Das et al. [2]	↓↓	Smaller heaters with 4.5 and 6-mm outside diameter	Al_2O_3–water nanofluid (4–16 wt%)
Bang and Chang [3]	↓↓	100-mm^2 surface	Al_2O_3–water nanofluid (4–16 wt%)
Shi M H et al. [7]	↑↓	1-kW film heater	Fe–water nanofluid (0.4–4 wt%) Al_2O_3–water nanofluid (0.4–8 wt%)
Vassallo et al. [8]	≈	NiCr wire heater	Silica–water nanofluid
Tu et al. [4]	↑↑	26 · 40-mm^2 rectangular surface	Al_2O_3–water nanofluid (1–8 wt%)
Witharana [5]	↑↑	100-mm diameter heater surface	Gold–water nanofluid (0.004 wt%)
Wen and Ding [6]	↑↑	3-mm SS disc heater	Al_2O_3–water nanofluid (0.32–1.25 vol%)

regarding the same. To clearly understand and explain boiling of nanofluids, more systematic experiments over wider ranges of parameters such as heating surface geometry, surface characteristics, particle properties, suspension thermophysical properties, and surfactants are required. A summary of pool boiling of nanofluids is presented in Table III.

XII. Studies on CHF in Pool Boiling

In another very interesting study, Vassallo et al. [187] observed that the CHF is very interestingly modified by nanofluids. A two to three times increase in CHF is seen by suspending SiO_2 nanoparticles in water as shown in Fig. 29. The results indicate that for NiG wire CHF for 30 nm as well as 100-nm silica particle suspension was substantially higher than the CHF limit suggested by Zuber [189]. The enhancement of CHF was further confirmed by

You *et al.* [186].The test results showed that the enhancement of CHF was drastic when nanofluid is used as a cooling liquid instead of pure water. The tested nanofluid contains Al_2O_3 nanoparticles dispersed in distilled and deionized water. $CHF_{nanofluids}$ was found to be about three times larger than the CHF_{water}. The average size of departing bubbles increased and the bubble frequency decreased significantly in nanofluids compared to those in the pure water. Later on, Bang and Chang [183] also observed similar phenomenon during their pool boiling studies. The measured pool boiling curves of nanofluids saturated at 60°C demonstrated that the CHF increased dramatically (200% increase) compared to pure water; however, the nucleate boiling heat transfer coefficients appeared to be about the same.

In yet another study, Bang and Chang [190] used different volume concentrations of alumina nanoparticles. The CHF has been enhanced in not only horizontal but also vertical pool boiling. The change of CHF was proposed to be due to a possible surface coating effect that would change the nucleation site density. The enhancement of the CHF observed is much lower than that observed in Vassallo *et al.* [187]. Possible reasons for this were suggested to be difference in the nanofluid used as well as the geometry of the heated surface. Bang and Chang [190] conducted experimental observation studies which confirmed the existence of a liquid film separating a vapor bubble from a heated solid surface using a nanofluid. Alumina nanofluid was used as a colored fluid in order to distinguish between the liquid phase and the vapor phase in a complex boiling environment. They showed that a liquid film under a massive vapor bubble adheres to a heated solid surface. It is observed that the liquid film comes into being trapped in a dynamic coalescence environment of nucleate bubbles, which grow and depart continuously from the heated surface. Some of the preparation methods need combination of physical and chemical methods like that used in [108] to prepare aqueous solution of CNTs. The preparation of nanofluid included ultrasonication of CNTs under dry conditions, dispersing the sonicated CNTs into a preset amount of distilled water containing gum arabic dispersant and adjusting the suspension to a preset pH level and high shear homogenization of the dispersion.

However, no study on boiling of metallic nanofluids as well as flow boiling of nanofluids is available so far.

XIII. Applications of Nanofluids

In the process of applying nanofluid for commercial cooling, Tzeng *et al.* [191] studied the effect of nanofluids when used as engine coolants. CuO

(4.4 wt%) and Al_2O_3 (4.4 wt%) nanoparticles and antifoam were mixed with automatic transmission oil for the study. A comparison was made between their heat transfer performance and that of oil without adding such substances. The experimental platform was a real-time four-wheel drive (4WD) transmission system. It adopts advanced rotary blade coupling (RBC), where a high local temperature occurs easily at high rotating speed. The experiment measured the axial temperature distribution of the RBC exterior at four different rotating speeds (400, 800, 1200, and 1,600 rpm). The experimental results showed that antifoam–oil had the highest temperature distribution in the same conditions and accordingly the worst heat transfer effect and CuO–oil has the lowest temperature distribution at both high and low rotating speeds and accordingly the best heat transfer effect.

Gosselin and Silva [172] explored the scope of optimizing particle fraction for maximizing the thermal performance of nanofluid flows under appropriate constraints. They argued that when the particles are few, the heat transfer rate that is achieved is small, whereas too many particles result in large shear stresses and pumping power requirement. This competition reveals a tradeoff opportunity for maximizing the heat transfer rate, at constant pumping power, by selecting the appropriate amount of particles. Laminar and turbulent boundary layer flows in forced and natural convections were considered. In forced convection, the power dissipation was constrained to highlight the competing effects of the thermal conductivity and viscosity variations due to the presence of the particles. In the study, the authors used Hamilton–Crosser model for predicting effective conductivity of the nanofluid which clearly underpredicts the anomalous enhancement in thermal conductivity which becomes the major draw back to the proposed optimization model. Moreover, experimental studies show that increase in pumping power due to addition of particles was not that significant.

Chein and Huang [192] considered silicon microchannel heat sink performance using nanofluids as coolants. The nanofluid was a mixture of pure water and nanoscale Cu particles with various volume fractions. Theoretical and experimental correlations for friction factor and Nusselt number variation were used to analyze the performance of the microchannel. The nanofluid was treated as a single-phase fluid with properties changed and the effect of thermal dispersion due to particle random motion is also included based on the experimental correlation for convective heat transfer in laminar flow by Li and Xuan [193]. Because of the increased thermal conductivity and thermal dispersion effects, it was found that the performances were greatly improved when nanofluids were used as the coolants. In addition, it was observed that the existence of nanoparticles in the fluid did not produce extra pressure drop because of small particle size and low particle volume fraction.

Koo and Kleinstreuer [194] investigated on the conjugate heat transfer problem for microheat sinks numerically considering two types of nanofluids, that is, CuO nanospheres at low volume concentrations in water and ethylene glycol. The effective thermal conductivity and dynamic viscosity of nanofluids were modeled which composed of two parts: the conventional static part and a dynamic part which originates from the particle Brownian motion. The effect of Brownian motion on the effective fluid viscosity was less significant than that on the effective thermal conductivity. The impact of nanoparticle concentrations in mixture flows on the microchannel pressure gradients, temperature profiles, and Nusselt numbers was computed, with respect to aspect ratio, viscous dissipation, and enhanced temperature effects. Based on the studies, it was suggested to use high Prandtl number carrier fluids and high thermal conductive nanoparticles for better performance of microchannel. It was also suggested that in order to minimize particle–particle and particle–wall interactions, particles with a dielectric constant close to that of the base fluid and a wall material is to be selected, such that particle–wall attraction is minimized.

On one hand, it is obvious from the above examples that nanofluids have the potential for application in variety of applications. On the other hand, it is obvious that the application-oriented research in nanofluids is at its very beginning. With more confirmation on fundamental behavior, research exploration in this area is in the offing.

XIV. Summary and Future Direction of Research

Ever since Choi [16] proposed abnormal enhancement of thermal conductivity in nanoparticle suspension (now popularly known as nanofluids), the interest in them is continuously on the rise. The past 3 years have seen an exponential growth in the number of research publications dealing with different aspects of synthesis, characterization, thermal conduction measurement, theorization, convection, and boiling studies on nanofluids. Considering the tremendous potential of using these fluids for new generation of cooling systems for microelectronic devices and energy-intensive optical (LASER) and optoelectronic (X-ray, fiber optics) devices, the number of studies in this area is likely to grow faster for which this chapter can be of help. This is particularly important considering the fact that the nature of research in this area is strongly interdisciplinary and the publications are spread over a large area of journals and conferences involving mechanical engineering, material science, chemical engineering, synthetic chemistry, physical chemistry, spectroscopy, electrical engineering, physics, computational sciences, micro fluidics, and biological sciences.

The first major conclusion that can be drawn from these studies is that the nanofluid has a very bright possibility for usage in cooling and related technology which can very easily overcome the usual problems with common slurries such as sedimentation, clogging, increased pressure drop, erosion, and applicability to microchannels. Nanofluids of ceramic and pure metallic particles have been produced by the conventional two-step method where the particles are first produced by methods such as IGC or chemical vapor deposition and then the particles are dispersed in the fluid using various methods such as physical dispersion and chemical dispersion methods where various techniques such as ultrasonic vibration, use of surfactants, or control of pH can be used. However, care has to be taken to use additives because they can change the properties of the fluids substantially and under certain conditions can even promote sedimentation. Compared to this, the one-step method which produces nanofluids directly without producing any dry nanoparticles seems to be promising. DEC, citrate reduction, or Brust method seems to be very efficient single-step method but the yield in these methods is small. Hence, the method of upscaling of these techniques is required to be investigated in future. Electrodynamic spraying system seems to have a good potential for large amount of continuous production of nanofluids. A large number of characterization techniques predominantly with electron microscopy have been reported. However, most of these studies are carried out with dried nanofluids; their behavior and particularly movements inside the fluids are yet to be characterized.

For measuring thermal conductivity of nanofluids, the very first need is to standardize the measurement techniques. There are few contrasting reports for thermal conductivity values of same configuration of nanofluids, measured by different research groups, sometimes by the same measurement technique and sometimes by different techniques. Various measurement techniques such as steady-state parallel plate, transient hot wire, temperature oscillation, 3-ω hot wire, optical beam deflection, and forced Rayleigh scattering have been used. Each of these techniques is having different accuracy as well as applicability. It is utmost important that the working principle of any technique is appropriate for the particular nanofluid used. In general, thermal conductivity of nanofluids is found to be much higher compared to what is expected from the classical Maxwell's theory or its various improved versions. The enhancement in thermal conductivity of oxide nanofluids is found to be relatively less, whereas the metallic nanofluids seem to enhance the thermal conductivity "anomalously." Also, the enhancement in thermal conductivity is found to be relatively higher at lower volume fractions, especially for metallic nanofluids, showing a nonlinear trend in the thermal conductivity enhancement of nanofluids with variation in volume fraction. Particle size reduction has a tremendous effect on the enhancement, suggesting almost an inverse relationship with the enhancement in thermal conductivity. An important effect on the enhancement of nanofluids comes from temperature. For a little rise in temperature, a very significant increase in the enhancement in thermal conductivity of nanofluids is observed. The maximum enhancement in thermal conductivity is observed with CNT suspensions. The high conductivity of CNTs and the large aspect ratio of them are proposed to be the reasons behind it. A comprehensive dataset of thermal conductivity of oxide, metallic, and CNT nanofluids with variation in particle volume fraction, particle and fluid properties, nanofluid temperature, particle size variation, same combination of nanofluid with different preparation techniques, and thermal conductivity measurement with different techniques is the need of the hour.

On the theoretical investigation of thermal conductivity of nanofluids, various mechanisms such as Brownian motion, ballistic heat transport, liquid layering, surface diffusion, and liquid clustering have been tried. There are quite a few studies considering same type of particle clustering analyzed using fractal dimensions. There are also liquid layering studies which assume liquid shell around the particles that behave like solids. However, the predictions are matched with experiments using adjustable parameters of shell thickness and shell conductivity which are questionable. Some of these models were able to model both oxide nanofluids and CNT nanofluids, which are commendable. One of the most promising mechanisms suggested is microconvection setting due to the Brownian motion of particles in liquids. As the particles move inside liquid, they help in reducing the temperature gradient in the liquid by acting as heat package carriers as well as by agitating the liquid. Various groups have treated this effect in different ways; however, the mechanism seems justifying all the major effects observed experimentally such as temperature effect, particle size effect, and particle volume fraction effect. Hence, the theoretical picture is still gray; in future, efforts may be concentrated particularly with attention toward particle movement, nanoconvection, and existence as well as nature of liquid layering. It is also important to use particle properties including the size effect rather than bulk properties in these models. It is important that any model to be developed in future be tested against a large number of data of ceramic, metallic, and nanotube-based nanofluids, and with respect to temperature and particle volume fraction, rather than the present practice of testing with limited range of measurement. Also if there are adjustable parameters in the model, then their values should be justified by the physics of the problem rather than by simple empirical treatment.

The studies on convective energy transport of nanofluids have just begun. The amount of experimental works in the convective heat transfer is still insufficient. One criticism on convective studies has been the use of electrical heating which may interfere with the motion of the particles with surface charge. Hence, purely thermal test with hot fluids may be used in future. Recently, researchers have tried to theoretically model the convective phenomenon. There are a number of studies in which the nanofluids are modeled as a single fluid with modified properties. Experimental studies clearly demonstrate that such a treatment is not permissible because other mechanisms such as thermophoresis, and thermal dispersions can play a significant role in convection of nanofluids. There are many issues to be dealt carefully while modeling convection such as thermal conductivity, Brownian motion of particles, particle migration, and variable property change with temperature. Also the majority of the convective studies have been made with oxide particles and it will be interesting to know the energy transport of low concentration nanofluids with metallic particles as well as with additional effects such as application of microwave. The future direction in the convective studies should be first with metallic nanoparticles in standard geometries to consider heat transfer enhancement, transition to turbulence, and hydraulic behavior which can further be extended in geometries such as microchannels where it may be of good prospect. The application-oriented research in nanofluids is in its infancy and is expected to grow at a faster rate in the foreseeable future which will define the future of nanofluids and its present promises.

Convection with phase change seems to be interesting. Boiling heat transfer seems to be adversely affected by plugging of nucleation sites. However, contradictory claims have come out that at lower particle concentration boiling is enhanced and in certain cases increase in roughness takes place. This area is still very foggy and a systematic study of the relative size of particles and surface cavities may hold the key. It is important to get to the mechanisms of boiling heat transfer in presence of the particles. In all cases, the CHF seems to be enormously enhanced due to the presence of the nanoparticles. The physical phenomenon behind this is unclear. Probably, wetting behavior of the surface in presence of the particles is a key area that needs investigation. This area holds tremendous promise for enhancing safety in power plants and boiling water nuclear reactors.

Thus, this chapter demonstrates the tremendous potential of nanofluids and the great enthusiasm developed within the research community thereof. However, for its thermal applications, there is some more way it has to travel before being seriously accepted as a candidate of future cooling technology. Future study needs to be concentrated on synthesis, stabilization, and application of metallic and CNT nanofluids. Comprehensive experimental data on conductive as well as convective studies and accurate predictive models for heat transfer behavior are the ultimate goals of the research in nanofluids. However, this has to be done always, keeping physics of the thermal transport mechanism in mind.

Nomenclature

ACRONYMS

A_p	particle surface area
CHF	critical heat flux
C_p	specific heat
C_D	Drag coefficient
c	cluster
D	channel hydraulic diameter (m)
d_{nano}	diameter of nanoparticle
F_D	drag force
G_o, G_L	phase shift of the input oscillation
g	gravity
h	enthalpy
I	turbulent intensity
J_μ	flux due to viscosity gradient
J_b	flux due to nonuniform shear
J_c	flux due to Brownian motion
k	turbulent kinetic energy
k_{eff}	effective thermal conductivity of slurry (W mK^{-1})
k_f	thermal conductivity of liquid (W mK^{-1})
k_p	thermal conductivity of solid particles (W mK^{-1})
k_b	Boltzman constant
$k_{d,r}$	dispersion coefficient in radial direction
$k_{d,x}$	dispersion coefficient in axial direction
L	channel length (m)
l	length of wire
n	shape factor (for sphere = 3, for cylinder = 6)
p	pressure
q	heat per unit length
R	tube radius
T	temperature
t	time
$\bar{u}_p$	Brownian motion of nanoparticles
V_p	particle volume
V_{dr}	drift velocity
w	pumping power
x	axial distance

GREEK SYMBOLS

ϕ	volume fraction of nanoparticles
λ	thermal conductivity (W mK^{-1})
μ_b	dynamic viscosity at bulk temperature (N s m^{-2})
μ_w	dynamic viscosity at wall temperature (N s m^{-2})
μ_l	liquid viscosity
μ_t	turbulent viscosity
ρ	density
ν	kinematic viscosity
α_p	diffusivity of particles
τ	shear stress
$\dot{\gamma}_r$	shear rate
β_T	temperature gradient
ε	dissipation rate

SUBSCRIPTS AND SUPERSCRIPTS

p	particle
f	fluid
BF	base fluid
nf	nanofluid

References

1. Feynman, R. (1959). Address given at annual meeting of American Physical Society, entitled "There's Plenty of Room at the Bottom: An Invitation to Enter a New Field of Physics" *Engineering and Science* magazine, vol. XXIII, no. 5, February 1960.
2. Rohrer, H. (1996). *Microelectronic Eng.* **32**(1–4), 5–14.
3. Duncan, A. B. and Peterson, G. P. (1994). *Appl. Mech. Rev.* **47**(9), 397–428.
4. Majumdar, A. (1998). Microscale energy transport in solids. In "Microscale Energy Transport" (C. L. Tien, A. Majumdar, and F. Gerner, eds.), Taylor & Francis, Washington, DC, USA.

5. Tuckerman, D. B. and Pease, R. F.W. (1981). *IEEE Electron. Device Lett.* **2**, 126–129.
6. Choi, S. U. S., Rogers, C. S., and Mills, D. M. (1992). High-performance microchannel heat exchanger for cooling high-heat-load x-ray optical elements. *In* "Micromechanical Systems" (D. Cho, J. P. Peterson, A. P. Pisano, and C. Friedrich, eds.), Vol. 40, p. 83. American Society of Mechanical Engineers, New York, NY, USA.
7. Kandlikar, S. G. (2002). *Exp. Thermal Fluid Sci.* **26**(2–4), 389–407.
8. Kandlikar, S. G. (2004). *J. Heat Transf.* **126**(1), 8–16.
9. Kandlikar, S. G. and Grande, W. J. (2002). *Heat Transf. Eng.* **25**(1), 3–17.
10. Bergles, A. E., Lienhard, J. H., Kendall, G. E., and Griffith, P. (2003). *Heat Transf. Eng.* **24**(1), 18–40.
11. Thome, J. R., Dupont, V., and Jacobi, A. M. (2004). *Int. J. Heat Mass Transf.* **47**(14–16), 3375–3385.
12. Maxwell, J. C. (1881). "A Treatise on Electricity and Magnetism." 2nd edn. Vol 1, Clarendon Press, Oxford, UK.
13. Hamilton, R. L. and Crosser, O. K. (1962). *Ind. Eng. Chem. Fundam.* **1**(3), 187–191.
14. Wasp, F. J. (1977). "Solid–Liquid Slurry Pipeline Transportation." Trans. Tech. Publication, Berlin, Germany.
15. Zhao, C. Y. and Lu, T. J. (2002). *Int. J Heat Mass Transf.* **45**(24), 4857–4869.
16. Choi, S. U. S. (1995). Enhancing thermal conductivity of fluids with nanoparticles. *In* "Developments and Applications of Non-Newtonian Flows" (D. A. Singer, and H. P. Wang, eds.), Vol. FED 231, pp. 99–105. ASME, New York, NY, USA.
17. Pozhar, L. A., Kontar, E. P., and Hua, M. Z.C. (2002). *J. Nanosci. Nanotechnol.* **2**(2), 209–227.
18. Pozhar, L. A. (2000). *Phys. Rev. E* **61**(2), 1432–1446.
19. Pozhar, L. A. and Gubbins, K. E. (1997). *Phys. Rev. E* **56**(5), 5367–5396.
20. Kim, P., Shi, L., Majumdar, A., and McEuen, P. L. (2001). *Phys. Rev. Lett.* **87**, 215502.
21. Das, S. K., Putra, N., Thiesen, P., and Roetzel, W. (2003). *ASME J. Heat Transf.* **125**, 567–574.
22. Lee, S., Choi, S. U. S., Li, S., and Eastman, J. A. (1999). *J. Heat Transf.* **121**, 280–289.
23. Xuan, Y. and Li, Q. (2000). *Int. J. Heat Fluid Flow* **21**, 58–64.
24. Das, S. K., Choi, S. U. S., and Patel, H. E. (2006). *Heat Transf. Eng.* **27**, 3–19.
25. Wang, X. Q. and Mujumdar, A. S. (2007). *Int. J. Thermal Sci.* **46**, 1–19.
26. Daungthongsuk, W. and Wongwises, S. (2007). *Renewable Sustain. Energy Rev.* **11**, 797–817.
27. Gandhi, K. S. (2007). *Curr. Sci.* **92**, 717–718.
28. Trisaksri, V. and Wongwises, S. (2007). *Renewable Sustain. Energy Rev.* **11**, 512–523.
29. Gleiter, H. (1989). *Prog. Mater. Sci.* **33**(4), 223–315.
30. Wang, X., Xu, X., and Choi, S. U. S. (1999). *J. Thermophys. Heat Transf.* **13**, 474–480.
31. Turker, M. (2004). *Mater. Sci. Eng. A* **367**(1–2), 74–81.
32. Akoh, H., Tsukasaki, Y., Yatsuya, S., and Tasaki, A. (1978). *J. Cryst. Growth* **45**, 495–500.
33. Eastman, J. A., Choi, S. U. S., Li, S., Yu, W., and Thomson, L. J. (2001). *Appl. Phys. Lett.* **78**(6), 718–720.
34. Eastman, J. A., Choi, S. U. S., Li, S., and Thompson, L. J. (1997). Enhanced thermal conductivity through the development of nanofluids. *In* "Proceedings of the Symposium on Nanophase and Nanocomposite Materials II." Vol. 457, pp. 3–11. Materials Research Society, Boston, MA, USA.
35. Eastman, J. A., Choi, S. U. S., Li, S., Soyez, G., Thompson, L. J., and DiMelfi, R. J. (1998). *J. Metastable Nanocrystal. Mater.* **2**, 629.
36. Xie, H., Wang, J., Xi, T., and Liu, Y. (2002). *Int. J. Thermophys.* **23**(2), 571–580.
37. Andrews, R., Jacques, D., Rao, A. M., Derbyshire, F., Qian, D., Fan, X., Dickey, E. C., and Chen, J. (1999). *Chem. Phys. Lett.* **303**, 467–474.

38. Patel, H. E., Das, S. K., Sundararajan, T., Sreekumaran, N. A. George, B., and Pradeep, T. (2003). *Appl. Phys. Lett.* **83**(14), 2931–2933.
39. Enustun, B. V. and Turkevich, J. (1963). *J. Am. Chem. Soc.*, **85**, 3317–3328.
40. Brust, M., Walker, M., Bethell D., Schiffrin, D. J., and Whyman, R. (1994). *J. Chem. Soc. Chem. Commun.* 801–802.
41. Zhu, H. T., Lin, Y. S., and Yin, Y. S. (2004). *J. Colloid Interface Sci.* **277**, 100–103.
42. Choi, S. U. S., Zhang, Z. G., Yu, W., Lockwood, F. E., and Grulke, E. A. (2001). *Appl. Phys. Lett.* **79**, 2252–54.
43. Platt, U. (1994). Differential optical absorption spectroscopy (DOAS). *In* "Air Monitoring by Spectroscopic Techniques" (M. Sigrist, ed.), pp. 27–84. John Wiley & Sons Inc., New York, NY, USA.
44. Xie, H. Q., Wang, J. C., Xi, T. G., Liu, Y., Ai, F., and Wu, Q. R. (2002). *J. Appl. Phys.* **91**(7), 4568–4572.
45. Aklesh Lakhtakia (2004). "Handbook of Nanotechnology: Nanometer Structure Theory, Modeling, and Simulation" http://members.asme.org/catalog/ItemView.cfm?ItemNumber=802159.
46. Hiemenz, P. C. and Rajagopalan, R. (1997). "Principles of Colloid and Surface Chemistry." 3rd edn. Marcel Dekker, New York, NY, USA.
47. Ross, S. and Morrison, I. D. (1988). "Colloidal Systems and Interfaces." Wiley, New York, NY, USA.
48. Xie, H., Wang, J., Xi, T., and Liu, Y. (2001). *J. Chin. Ceramic Soc.* **29**(4), 361–364.
49. Dejaguin, B. V. and Landau, L. D. (1941). *Acta Physiochim. USSR* **14**, 633–652.
50. Verwey, E. J.W. and Overbeek, J. T.G. (1948). "Theory of Stability of Lyophobic Colloids." Elsevier, Amsterdam, The Netherlands.
51. Reerink, H. and Overbeek, J. T.G. (1954). *Discuss. Faraday Soc.* **18**, 74–84.
52. Sugimoto, T., ed. (2000). "Fine Particles: Synthesis, Characterization, and Mechanism of Growth." *Surfactant Science Series*, Vol. 92, pp. 290–299. Marcel Dekker, New York, NY, USA.
53. Chen, D. R. and Kaufman, S. L. (1995). *J. Aerosol. Sci.* **26**, 963–977.
54. Chen, D. R. and Pui, D. Y.H. (1997). *Aerosol. Sci. Technol.* **27**, 367–380.
55. Maxwell-Garnett, J. C. (1904) *Philos. Trans. R. Soc. A* **203**, 385–420.
56. Jeffrey, D. J. (1973). *Proc. R. Soc. Lond. Ser. A* **335**(1602), 355–367
57. Davis, R. H. (1986). *Int. J. Thermophys.* **7**(3), 609–620.
58. Lu, S. and Lin, H. (1996). *J. Appl. Phys.* **79**, 6761–6769.
59. Murshed, S. M.S., Leong, K. C., and Yang, C. (2005). *Int. J. Thermal Sci.* **44**, 367–373.
60. Zhu, H., Zhang, C., Liu, S., Tang, Y., and Yin, Y. (2006). *Appl. Phys. Lett.* **89**, 023123.
61. Hwang, Y., Lee, J. K., Lee, C. H., Jung, Y. M., Cheonga, S. I., Lee, C. G., Ku, B. C., and Jang, S. P. (2007). *Thermochimica Acta* **455**, 70–74.
62. Kim, S. H., Choi, S. R., and Kim, D. (2007). *ASME J. Heat Transf.* **129**, 298–307.
63. Carslaw, H. S. and Jaeger, J. C. (1959). "Conduction of Heat in Solids." 2nd edn. Oxford University Press, New York, NY, USA.
64. Liu, M. S., Lin, M. C.-C., Tsai C. Y., and Wang, C.-C. (2006). *Int. J. Heat Mass Transf.* **49**, 3028–3033.
65. Xie, H. Q., Wang, J. C., Xi, T. G., Liu, Y., and Ai, F. (2002). *J. Mater. Sci. Lett.* **21**, 1469–1471.
66. Hong, T. K., Yang, H. S., and Choi, C. J. (2005). *J. Appl. Phys.* **97**, 064311.
67. Hong, K. S., Hong, T.-K., and Yang, H.-S. (2006). *Appl. Phys. Lett.* **88**, 031901.
68. Murshed, S. M.S., Leong, K. C., and Yang, C. (2006). *J. Phys. D Appl. Phys.* **39**, 5316–5322.
69. Chopkar, M., Das, P. K., and Manna, I. (2006). *Scr. Mater.* **55**, 549–552.
70. Biercuk, M. J., Llaguno, M. C., Radosavljevic, M., Hyun, J. K., Johnson, A. T., and Fischer, J. E. (2002). *Appl. Phys. Lett.* **80**, 2767–2772.

71. Xie, H., Lee, H., Youn, W., and Choi, M. (2003). *J. Appl. Phys.* **94**, 4967–4971.
72. Assael, M. J., Metaxa, I. N., Arvanitidis, J., Christophilos, D., and Lioutas, C. (2005). *Int. J. Thermophys.* **26**, 647–664.
73. Assael, M. J., Metaxa, I. N., Kakosimos, K., and Constantinou, D. (2006). *Int. J. Thermophys.* **27**(4), 999–1017.
74. Hwang, J., Ahn, Y. C., Shin, H. S., Lee, C. G., Kim, G. T., Park, H. S., and Lee, J. K. (2005). *Curr. Appl. Phys.* **6**(6), 1068–1071.
75. Liu, M. S., Lin, M. C.C., Haung, I. T., and Wang, C. C. (2005). *Int. J. Heat Mass Transf.* **32**, 1202–1210.
76. Hong, H., Wright, B., Wensel, J., Jin, S., Ye, X., and Roy, W. (2007). *Synth. Met.* **157**(10-12). 437–440.
77. Zhu, H., Zhang, C., Tang, Y., Wang, J., Ren, B., and Yin, Y. (2007). *Letters to the Editor, Carbon* **45**, 203–228.
78. Yang, Y., Grulkea, E. A., Zhang, Z. G., and Wu, G. (2006). *J. Appl. Phys.* **99**, 114307.
79. Shaikh, S., Lafdi, K., and Ponnappan, R. (2007). *J. Appl. Phys.* **101**, 064302.
80. Czarnetzki, W. and Roetzel, W. (1995). *Int. J. Thermophys.* **16**(2), 413–422.
81. Chon, C. H. and Kihm, K. D. (2005). *Tans ASME J. Heat Transf.* **127**, 810.
82. Li, C. H. and Peterson, G. P. (2006). *J. Appl. Phys.* **99**, 084314.
83. Murshed, S. M. S., Leong, K. C., and Yang, C. (2008). *Int. J. Thermal Sci.* **47**(5), 560–568.
84. Li, C. H. and Peterson, G. P. (2007). *J. Appl. Phys.* **101**, 044312.
85. Yang, B. and Han, Z. H. (2006). *Appl. Phys. Lett.* **88**, 261914.
86. Zhang, X., Gu, H., and Fujii, M. (2007). *Exp. Thermal Fluid Sci.* **31**, 593–599.
87. Putnam, S. A., Cahill, D. G., Braun, P. V., Ge, Z., and Shimmin, R. G. (2006). *J. Appl. Phys.* **99**, 084308.
88. Venerus, D. C., Kabadi, M. S., Lee, S., and Perez-Luna, V. (2006). *J. Appl. Phys.* **100**, 094310.
89. Bruggeman, D. A.G. (1935). *Annalen der Physik. Leipzig* **24**, 636–679.
90. Bonnecaze, R. T. and Brady, J. F. (1990). *Proc. R. Soc. Lond. A* **430**, 285–313.
91. Bonnecaze, R. T. and Brady, J. F. (1991). *Proc. R. Soc. Lond. A* **432**, 445–465.
92. Cheng, S. C. and Vachon, R. I. (1969). *Int. J. Heat Mass Transf.* **12**, 249.
93. Yu, W. and Choi, S. U. S. (2003). *J. Nanoparticles Res.* **5**(1–2), 167–171.
94. Hirtzel, C. S. and Rajagopalan, R. (1985). "Colloidal Phenomena." Noyes Publications, Park Ridge, NJ, USA.
95. Keblinski, P., Phillpot, S. R., Choi, S. U. S., and Eastman, J. A. (2002). *Int. J. Heat Mass Transf.* **45**, 855–863.
96. Chen, G. (1996). *J. Heat Transf. Trans. ASME* **118**(3), 539–545.
97. Wang, B., Zhou, L., and Peng, X. (2003). *Int. J. Heat Mass Transf.* **46**, 2665–2672.
98. Pitchumani, R. and Yao, S. C. (1991). *J. Heat Transf.* **113**, 788–796.
99. Ma, K. Q. and Liu, J. (2007). *Phys. Lett. A* **361**, 252–256.
100. Xue, Q. Z. (2003). *Phys. Lett. A* **307**, 313–317.
101. Xue, Q. Z. (2005). *Physica B* **368**, 302–307.
102. Xue, Q. and Xu, W. M. (2005). *Mater. Chem. Phys.* **90**, 298–301.
103. Schwartz, L. M., Garboczi, E. J., and Bentz, D. P. (1995). *J. Appl. Phys.* **78**(10), 5898–5908.
104. Xie, H., Fujii, M., and Zhang, X. (2005). *Int. J. Heat Mass Transf.* **48**, 2926–2932.
105. Yu, C. J., Richter, A. G., Datta, A., Durbin, M. K., and Dutta, P. (2000). *Physica B* **283**, 27–31.
106. Xue, L., Keblinski, P., Phillpot, S. R., Choi, S. U. S., and Eastman, J. A. (2004). *Int. J. Heat Mass Transf.* **47**, 4277–4284.
107. Leong, K. C., Yang, C., and Murshed, S. M. S. (2006). *J. Nanoparticle Res.* **8**, 245–254.

108. Prasher, R., Evans, W., Meakin, P., Fish, J., Phelan, P., and Keblinski, P. (2006). *Appl. Phys. Lett.* **89**, 143119.
109. Feng, Y., Yu, B., Xu, P., and Zou, M. (2007). *J. Phys. D Appl. Phys.* **40**, 3164–3171.
110. Wojnar, R. (2001). *Acta Phys. Pol. B* **32**(2), 333–349.
111. Gitterman, M. (1995). *Phys. Rev. E* **52**(1), 303–306.
112. Gupta, A. S. (1999). *Heat Mass Transf.* **35**(4), 315–320.
113. Xuan, Y., Li, Q., and Hu, W. (2003). *AIChE J.* **49**(4), 1038–1043.
114. Jang, S. P. and Choi, S. U. S. (2004). *Appl. Phys. Lett.* **84**(21), 4316–4318.
115. Xu, J., Yu, B., Zou, M., and Xu, P. (2006). *J. Phys. D Appl. Phys.* **39**, 4486–4490.
116. Hemanth, K. D., Patel, H. E., Rajeev, K. V. R., Sundararajan, T., Pradeep, T., and Das, S. K. (2004). *Phys. Rev. Lett.* **93**(14), 144301-1–144301-4.
117. Ren, Y., Xie, H., and Cai, A. (2005). *J. Phys. D Appl. Phys.* **38**, 3958–3961.
118. Prasher, R., Bhattacharya, P., and Phelan, P. E. (2005). *Phy. Rev. Lett.* **94**, 025901.
119. Prakash, M. and Giannelis, E. P. (2007). *J. Comput. Aided Mater. Des.* **14**, 109–117.
120. Patel, H. E., Sundararajan, T., Pradeep, T., Dasgupta, A., Dasgupta, N., and Das, S. K. (2005). *Pramana J. Phys.* **65**, 863–869.
121. Xuan, Y., Li, Q., Zhang, X., and Fujii, M. (2006). *J. Appl. Phys.* **100**, 043507
122. Patel, H. E., Sundararajan, T., and Das, S. K. (2008). *J. Nanoparticle Res.* **10**(1), 87–97.
123. Chon, C. H., Kihm, K. D., Lee, S. P., and Choi, S. U. S. (2005). *Appl. Phys. Lett.* **87**, 153107.
124. Patel, H. E., Sundararajan, T. and Das, S. K. (2008). *Pramana-J. Physics* **65**(5), 863–869.
125. Zhou, L. P. and Wang, B. X. (2002). Experimental research on the thermophysical properties of nanoparticle suspensions using the quasi-steady state method (in Chinese). In "Annual Proceedings of the Chinese Engineering Thermophysics, Shanghai, China." pp. 889–892.
126. Bhattacharya, P., Saha, S. K., Yadav, A., Phelan, P. E., and Prasher, R. S. (2004). *J. Appl. Phys.* **95**(11), 6492–6494.
127. Xuan, Y. and Yao, Z. (2005) *J. Heat Mass Transf.* **43**(3), 199–205.
128. Eapen, J., Li, J., and Yip, S. (2007). *Phys. Rev. Lett.* **98**, 028302.
129. Li, C. H. and Peterson, G. P. (2007). *Int. J. Heat Mass Transf.* **50**(23-24). 4668–4677.
130. Evans, W., Fish, J., and Keblinskia, P. (2006). *Appl. Phys. Lett.* **88**, 093116.
131. Nan, C. W., Shi, Z., and Lin, Y. (2003). *Chem. Phys. Lett.* **375**, 666–669.
132. Nan, C. W., Liu, G., Lin, Y., and Li, M. (2004). *Appl. Phys. Lett.* **85**, 3549–3551.
133. Xue, Q. Z. (2006). *Nanotechnology* **17**, 1655–1660.
134. Gao, L. and Zhou, X. F. (2006). *Phys. Lett. A* **348**, 355–360.
135. Ju, S. and Li, Z. Y. (2006). *Phys. Lett. A* **353**, 194–197.
136. Straussa, M. T. and Pober, R. L. (2006). *J. Appl. Phys.* **100**, 084328.
137. Pak, B. and Cho, Y. I. (1998). *Exp. Heat Transf.* **11**, 151–170.
138. Batchelor, G. K. (1977). *J. Fluid Mech.* **83**(1), 97–117.
139. Das, S. K., Putra, N., and Roetzel, W. (2003). *Int. J. Heat Mass Transf.* **46**(5), 851–862.
140. Ding, Y., Alias, H., Wen, D., and Williams, A. R. (2006). *Int. J. Heat Mass Transf.* **49**, 240–250.
141. Dittus, F. W. and Boelter, L. M. K. (1930). "Publication in Engineering." Vol. 2, p. 433. University of California, Berkeley, CA, USA.
142. Xuan, Y. and Li, Q. (2003). *J. Heat Trans.* **125**(1), 151–155.
143. Wen, D. and Ding, Y. (2004). *Int. J. Heat Mass Transf.* **47**, 5181–5188.
144. Shah, R. K. (1975). Thermal entry length solutions for the circular tube and parallel plates. In "Proceedings of the 3rd National Heat Mass Transfer Conference." Vol. 1. Paper No. HMT-11-75. Indian Institute of Technology, Bombay, India.

145. Yang, Y. Z., Zhang, G., Grulke, E. A., Anderson, W. B., and Wu, G. (2005). *Int. J. Heat Mass Transf.* **48**, 1107–1116.
146. Kabelac, S. and Kuhnke, J. F. (2006). Heat transfer mechanisms in nanofluids – Experiments and theory. *In* "Keynote Lecture – 13th International Heat Transfer Conferences, Sydney, Australia, August 13–18, 2006".
147. Sieder, E. N. and Tate, G. E. (1936). *Ind. Eng. Chem.* **28**, 1429–1435.
148. Oliver, D. R. (1962). *Chem. Eng. Sci.* **17**, 335–350.
149. Eubank, C. C. and Proctor, W. S. (1951). Effect of natural convection on heat transfer with laminar flow in tubes, M.Sc. Thesis in Chemical Engineering, Massachusetts Institute of Technology, Cambridge, MA, USA.
150. Heris, S. Z., Etemad, S. G., and Esfahany, M. N. (2006). *Int. Commun. Heat Mass Transf.* **33**(4), 529–535.
151. Nguyen, C. T., Roy, G., Gauthier, C., and Galanis, N. (2007). *Appl. Thermal Eng.* **27**(8–9), 1501–1506
152. Xuan, Y. and Roetzel, W. (2000). *Int. J. Heat Mass Transf* **43**, 3701–3707.
153. Taylor, G. (1954). *Proc. R. Soc. A* **225**, 473–477.
154. Aris, R. (1956). *Proc. R. Soc. Lond. Ser. A*, **235**, 67–77.
155. Dankwerts, P. V. (1953). *Chem. Eng. Sci.* **2**, 1–10.
156. Kaviany, M. (1991). "Principles of Heat Transfer in Porous Media." Springer-Verlag, New York, NY, USA.
157. Kaviany, M. (1994). "Convective Heat Transfer." Springer-Verlag, New York, NY, USA.
158. Roetzel, W., Luo, X., and Xuan, Y. (1993). *Exp. Fluid Sci.* **7**, 345–353.
159. Beckman, L. V., Law, V. J., Bailey, R. V., and von Rosenberg, D. U. (1990). *AIChE J.* **36**, 598–604.
160. Ding, Y. and Wen, D. (2005). *Powder Technol.* **149**, 84–92.
161. Phillips, R. J., Armstrong, R. C., Brown, R. A., Graham, A. L., and Abbott, J. R. (1992). *Phys. Fluids A* **4**, 30–40.
162. Buongiorno, J. (2006). *J. Heat Transf.* **128**, 240.
163. Bird, R. B., Stewart, W. E., and Lightfoot, E. N. (1960). "Transport Phenomena." Wiley & Sons, New York, NY, USA.
164. Maiga, S. E., Nguyen, C. T., Galanis, N., and Roy, G. (2004). *Superlattices and Microstructures* **35**, 543–557.
165. Roy, G., Nguyen, C. T., and Lajoie, P. (2004). *Superlattices Microstruct.* **35**(3–6), 497–511.
166. Palm, S. J., Roy, G., and Nguyen, C. T. (2006). *Appl. Thermal Eng.* **26**(17–18), 2209–2218.
167. Putra, N., Roetzel, W., and Das, S. K. (2003). *Heat Mass Transf.* **39**, 775–784.
168. Behzadmehr, A., Saffar-Avval, M., and Galanis, N. (2007) *Int. J. Heat Fluid Flow* **28**(2), 211–219.
169. Patankar, S. V. (1980). "Numerical Heat Transfer and Fluid Flow." Hemisphere, New York, NY, USA.
170. Mansour, R. B., Galanis, N., and Nguyen, C. T. (2007). *Appl. Thermal Eng.* **27**(1), 240–249.
171. Brinkman, H. C. (1952). *J. Chem. Phys.* **20**, 571–581.
172. Gosselin, L. and Da Silva, A. K. (2004). *Appl. Phys. Lett.* **85**(18), 4160–4162.
173. Wen, D. and Ding, Y. (2005). *Int. J. Heat Fluid Flow* **26**, 855–864.
174. Khanafer, K., Vafai, K., and Lightstone, M. (2003). *Int. J. Heat Mass Transf.* **46**, 3639–3653.
175. Jou, R. and Tzeng, S. (2006). *Int. Commun. Heat Mass Transf.* **33**(6), 727–736.
176. Kim, J., Kang, Y. T., and Choi, C. K. (2004). *Phys. Fluids* **16**(7), 2395–2401.
177. Kim, J., Kang, Y. T., and Choi, C. K. (2006). *Int. J. Refrigeration* **30**(2), 323–328.
178. Tiwari, R.K. and Das, M. K. (2007). *Int. J. Heat Mass Transf.* **50**(9–10), 2002–2018.

179. Polidori, G., Fohanno, S., and Nguyen, C. T. (2007). *Int. J. Thermal Sci.* **46**(8), 739–744.
180. Akbarinia, A. (2008). *Int. J. Heat Fluid Flow* **29**(1), 229–241.
181. Yang, Y. M. and Maa, J. R. (1983). *J. Heat Transf.* **105**, 190–192.
182. Das, S. K., Putra, N., and Roetzel, W. (2003b). *Int. J. Multiphase Flow* **29**, 1237–1247.
183. Bang, I. C. and Chang, S. H. (2005a). *Int. J. Heat Mass Transf.* **48**, 2407–2419.
184. Tsai, C. Y., Chien, H. T., Ding, P. P., Chang, B., Luh, T. Y., and Chen, P. H. (2004). *Mater. Lett.* **58**(9). 1461–1465.
185. Wen, D. and Ding, Y. (2005) *J Nanoparticle Res.* **7**, 265–274.
186. You, S. M., Kim, J. H., and Kim, K. M. (2003). *Appl. Phys. Lett.* **83**(16), 3374–3376.
187. Vassallo, P., Kumar, R., and D'Amico, S. (2004). *Int. J. Heat Mass Transf.* **47**(2), 407–411.
188. Witharana, S. (2003). Boiling of Refrigerants on Enhanced Surfaces and Boiling of Nanofluids, Ph.D. Thesis, Royal Institute of Technology, Stockholm, Sweden.
189. Zuber, N. (1958). *Trans. ASME* **80**, 711.
190. Bang, I. C. and Chang, S. H. (2005) *Appl. Phys. Lett.* **86**(13), 134107-1–134107-3.
191. Tzeng, S. C., Lin, C. W., and Huang, K. D. (2005). *Acta Mechanica* **179**, 11–23.
192. Chein, R. and Huang, G. (2005). *Appl. Thermal Eng.* **25**, 3104–3114.
193. Li, Q. and Xuan, Y. (2002). *Sci. China Ser. E* **45**, 408–416.
194. Koo, J. and Kleinstreuer, C. (2005). *Int. J. Heat Mass Transf.* **48**, 2652–2661.

AUTHOR INDEX

A

Adkins, D. R., 8, 10
Akbarinia, A., 175
Anand, S., 16, 18, 28, 34
Andraka, C. E., 10
Aris, R., 141
Assael, M. J., 106

B

Babin, B. R., 7–8, 11, 12, 27, 28, 31
Baker, K. W., 10
Bang, I. C., 179, 181, 182, 183, 184, 185
Basu, S., 65
Batchelor, G. K., 126, 127, 148
Beavers, G., 64–65
Bergles, A. E., 82
Berre, M. L., 51
Bhattacharya, P., 124
Biercuk, M. J., 105
Bird, R. B., 151
Boelter, L. M. K., 129–130, 154
Bologa, M. K., 31
Boman, B. L., 10
Bonnecaze, R. T., 114
Brennan, P. J., 67
Brinkman, H. C., 162, 168
Bruggeman, D. A. G., 114
Brust, M., 88, 94
Bryan, J. E., 31
Buongiorno, J., 151

C

Cao, Y., 2, 10
Carey, V. P., 65
Carslaw, H. S., 102
Catton, I., 15
Chang, S. H., 181, 182, 183, 184, 185
Chang, W. S., 15, 33
Chein, R., 186

Chen, D. R., 94
Chen, G., 117
Chen, Y. P., 48
Cheng, P., 48, 49
Chi, S. W., 2, 11, 17, 67
Chisholm, D., 67
Cho, Y. I., 126, 128, 129, 134, 154, 162
Choi, S. U. S., 82, 84–85, 90, 92, 104, 106, 114, 115, 118, 121, 163, 187
Chon, C. H., 112, 124
Chopkar, M., 104
Collier, J. G., 15, 33
Colwell, G. T., 15, 33
Cotter, T. P., 5, 11, 12, 67
Crosser, O. K., 84, 96, 97, 98, 100, 111, 113, 114, 121, 162, 165, 186
Czarnetzki, W., 107

D

da Silva, A. K., 162
Das, M. K., 173
Das, S. K., 85, 86, 90, 93, 100, 107, 112, 113, 114, 127, 176, 179, 180, 181, 182, 183, 184
Daungthongsuk, W., 86
Davis, R. H., 99
Debs, R. J., 31
Ding, Y., 128, 130, 131, 134, 135, 145, 146, 148, 151, 153, 166, 167, 182, 183, 184
Dittus, F. W., 129–130, 154
Do, K. H., 17, 29
Dresselhaus, M. S., 69
Duncan, A. B., 82
Dunn, P. D., 2, 11, 53, 67

E

Eapen, J., 124–125
Elias, T. I., 10
Eubank, C. C., 132
Evans, W., 125

F

Faghri, A., 2, 10, 12, 13, 53
Feldman, K. T., 67
Feng, Y., 119
Feynman, R., 81
Fletcher, L. S., 9

G

Galanis, N., 162
Gandhi, K. S., 86
Gao, L., 125
Gaugler, R., 2
Gay, F. W., 2
Gerner, F. M., 12, 50
Gleiter, H., 87
Gosselin, L., 162, 163, 186
Gottschlich, J. M., 10
Grande, W. J., 82
Groll, M., 53
Grover, G. M., 2
Grover, G., 2

H

Ha, J. M., 14, 16
Hamilton, R. L., 84, 96, 97, 98, 100, 111, 113, 114, 121, 162, 165, 186
Han, Z. H., 113
Heine, D., 53
Hemanth, K. D., 121
Heris, S. Z., 133, 134
Hiemenz, P. C., 122
Hoda, N., 18, 22, 32, 34–35, 59
Hong, H., 106
Hong, T. K., 104, 124
Huang, G., 186
Hwang, J., 106
Hwang, Y., 100, 106

I

Itoh, A., 2

J

Jaeger, J. C., 102
Jang, S. P., 121
Jeffrey, D. J., 99
Jiang, P., 65
Jones, T. B., 31
Josef, D. D., 64–65
Joshi, Y. K., 16
Jou, R., 168, 170
Ju, S., 125

K

Kabelac, S., 131, 134
Kalahasti, S., 16
Kandlikar, S. G., 82
Kang, S. W., 48, 50, 52
Kaviany, M., 141, 142
Keblinski, P., 115, 116
Khanafer, K., 167, 168
Khandekar, S., 17
Khrustalev, D., 12, 13
Kim, B. H., 65–66
Kim, J., 168–170, 171
Kim, S. H., 100
Kim, S. J., 16, 29
King, C. R., 2
Kleinstreuer, C., 187
Kojima, Y., 8
Koo, J., 187
Kroliczek, E. J., 67
Ku, B. C
Kumar, P., 18, 52, 56, 59, 70

L

Lee, M., 46, 51, 52
Lee, S., 90, 95, 96, 97, 103, 107, 109, 114, 127
Leong, K. C., 119
Li, C. H., 112, 113, 125
Li, Q., 88, 92–93, 94, 100, 129, 133, 134, 154, 186
Li, Z. Y., 125
Lightfoot, E. N., 151
Lin, H., 99
Liu, M. S., 103, 106

Loehrke, R. I., 31
Longtin, J. P., 12, 57
Lu, S., 99

M

Ma, H. B., 13, 14, 43, 48, 56
Ma, K. Q., 117
Maa, J. R., 175, 184
Maiga, S. E., 154, 155, 157
Majumdar, A., 82
Mallik, A. K., 7, 50
Mansour, R. B., 162, 165
Maxwell, J. C., 84, 96, 114–115
Maxwell-Garnett, J. C., 96, 117, 125
Meininger, M., 10
Melcher, J. R., 31
Mishiro, H
Moon, S. H., 29, 30, 51
Mujumdar, A. S., 86
Murakami, M., 27
Murshed, S. M. S., 100, 104, 107, 112–113

N

Nan, C. W., 125
Nguyen, C. T., 136
North, M. T., 8

O

Oliver, D. R., 132

P

Pak, B., 126, 128, 129, 134, 154, 162
Palm, S. J., 157, 158
Park, Y. S., 17
Patel, H. E., 88, 90, 94, 95, 101, 103, 111, 123, 124
Pease, R. F. W., 7
Perkins, L. P., 2
Perry, M. P., 31
Peterson, G. P., 2, 8–9, 13, 14, 16, 33, 34, 43, 44, 48, 56, 65–66, 68
Phillips, R. J., 145
Pitchumani, R., 117
PolderPolidori, G., 173
Prakash, M., 123

Prasher, R., 119, 123
Proctor, W. S., 132
Provotorov, V. P. J., 68
Putnam, S. A., 113
Putra, N., 157, 165, 166

R

Rajagopalan, R., 122
Reay, D. A., 2, 11, 53, 67
Ren, Y., 123
Riabov, V. V., 68
Riffat, S. B., 15
Roetzel, W., 107, 140, 141, 142, 143, 145, 163
Rohrer, H., 81
Rosenfeld, J. H., 8
Roy, G., 157

S

Sartre, V., 14
Sato, K., 27
Sato, M., 31
Savin, I. K., 31
Schneider, G. E., 17
Schwartz, L. M., 118
Shah, R. K., 16, 17, 131
Shaikh, S., 107
Sheu, T. S., 18
Shyu, R. J., 2
Sieder, E. N., 132, 133
Sobhan, C. B., 33
Stewart, W. E., 151
Stroes, G. R., 15
Sugumar, D., 53–54
Suh, J. S., 17, 49
Suman, B., 1–72

T

Tate, G. E., 132, 133
Taylor, G., 141
Terpstra, M., 2
Thome, J. R., 82
Tien, C. L., 67
Tio, K. K., 15
Tiwari, R. K., 173
Trefethen, L., 2
Trisaksri, V., 86

Tsai, C. Y., 182
Tuckerman, D. B., 7
Tzeng, S. C., 185
Tzeng, S., 168, 170

V

Van Veen, J. G., 2
Vassallo, P., 184, 185
Venerus, D. C., 113

W

Wang, B. X., 124
Wang, B., 117
Wang, C. Y., 13
Wang, X. Q., 86, 99
Wang, X., 99, 114, 115, 116, 163
Wasp, F. J., 84, 114, 168
Wayner, P. C., Jr., 65, 70
Weichold, M. H., 51
Wen, D., 130, 131, 134, 135, 145, 146, 148, 151, 166, 167, 182, 183
White, F. M., 48
Witharana, S., 183
Wong, H., 16
Wongwises, S., 86
Wu, D., 13, 33, 41, 42
Wu, H. Y., 49

X

Xie, H., 93, 100, 105, 118
Xie, H. Q., 90, 99, 103
Xu, J., 121
Xuan, Y., 88, 92–93, 94, 100, 120, 124, 129, 133, 134, 140, 141, 142, 143, 145, 154, 162, 163, 186
Xue, L., 119
Xue, Q. Z., 117, 118, 125

Y

Yang, B., 113
Yang, Y. M., 175
Yang, Y., 106
Yang, Y. Z., 131, 132, 133, 134
Yao, S. C., 124
You, S. M., 185
Yu, W., 114, 115, 118, 120, 163
Yu, Z. Q., 31, 32

Z

Zhang, J., 16, 30
Zhang, X., 113
Zhou, J., 30
Zhou, L. P., 124
Zhu, H. T., 88, 95
Zhu, H., 100, 106
Zuber, N., 184

SUBJECT INDEX

A

Accommodation theory, 15
Agglomeration, 85–7, 92–4, 103, 105, 149, 183
Air as coolant, 82–3
Aluminium oxide–water nanofluid
  enhancement in
    conductivity, 118
    thermal conductivity with temperature, 110
  pool boiling characteristics of, 182
Anand numerical model, 16
  See also Steady-state models
Applied Mechanics Reviews, 2
Argonne National Laboratory (ANL), 84–5

B

Babin model for trapezoidal MHP, 11
  See also Steady-state models
Bachelors formula, 148–9
  See also Convection in nanofluids
Batchelor model, 126–7
BMGN model, 163
Boiling in nanofluids
  pool boiling heat transfer at higher solid particle concentrations, 176
    bubble sliding mechanism, 176–7
    curve for, 185
    deterioration in, 176
    dimensionless boiling characteristics, 180
    fouling effect, 182
    heat flux, 178
    $Nu-Re_b$ characteristics, 179
    performance of, 178
    plugging effect, 182
    Rohsenow correlation, 181
    surface characteristics, 176
  pool boiling heat transfer at lower solid particle concentrations
    enhancement of, 182
    heat flux, 182
    heat transfer characteristics, 183

Boltzmann transport equation
  with nanoparticles in host medium, 117
Bond number, 6
Boussinesq approximation, 173
Brinkman model, 162
  for viscosity, 168
Brooks field viscometer for shear thinning behavior, 126
Brownian motion in nanofluids, 120
  Brownian diffusion, 151
  convection-like effects at nanoscale, 121
  driven microconvection around nanoparticles, 124
Bruggeman model, 114
Brunauer–Emmett–Teller method, 90
Brust method, 88–9, 94
  See also Synthesis of nanofluids
BTE. *See* Boltzmann transport equation

C

Cancer treatment, MHP catheter use in, 9, 68
Capillary limit for MHP, 55–6
  and critical heat input, 56
    analytical expression of, 56–7
    apex angle of V-shaped MHP, 58–9
    numerical method for calculation of, 56
    variation with inclination, 58
    variation with length, 58
  and dry-out length, 59
    analytical expression, 59–61
    numerical method, 59
    variation with heat input, 61
  mathematical model for, 56
Carbon nanotubes (CNTs), 87, 105–107
  CNT-based nanofluids
    conductivity, enhancement of, 118
    heat transfer coefficient, enhancement of, 135
    linear shear thinning behavior, 128
    study on convection with, 134
    thermal conductivity of, 134
    viscosity, 128

Chemical dispersion technique for nanofluids, 93–4
Citrate reduction method, 88
  See also Synthesis of nanofluids
Conjugate heat transfer problem for microheat sinks and nanofluids, 187
Constrained vapor bubble, 65
Convection in nanofluids
  analytical and numerical studies
    average heat transfer coefficient, 158
    Bachelors formula, 148–9
    BMGN model, 163
    boundary conditions, 143
    Brinkman model, 162
    conduction flux, 142
    continuity equation for, 151–4
    convection equations, 154–5
    Dankwert's boundary condition, 144
    dissipation rate, 161
    drift velocity, 160
    effective properties, 155
    friction factor, 145
    GdS model, 162
    heat and mass flow rate, 164
    laminar and turbulent forced flow, 164
    laminar convection in, 156
    mixture density and viscosity, 160
    momentum balance in control volume, 147
    Nusselt number, 144–5
    particle migration effect, 145
    primary phase velocity, 160
    procedure of Kaviany, 141
    shear relation, 160–1
    shear stress and rate, 147, 156
    SIMPLE algorithm, 162
    temperature-dependent properties, 157
    theorizing concept, 140
    thermal dispersion coefficient, 141–2
    turbulent kinetic energy, 160–1
    volume-averaged temperature and velocity vectors, 141
  experimental works on, 128
    convective heat transfer applications, 129
    entry length effect, 133
    laminar flow, 130
    local heat transfer coefficient for, 130–1
    Oliver correlation, 132
    regression analysis, 129
    Sieder Tate correlation, 132
    theory of thermal dispersion and, 130
    thermal entry length with constant heat flow, 131
    traditional turbulent flow equations, 134
  forced convection in, 126–8
  mechanisms involved in suspensions, 136–41
Cooling technology, need for development, 81–2
Copper oxide–water nanofluid
  enhancement in thermal conductivity with temperature, 110
Cotter model of small MHPs, 11–12
  See also Steady-state models
Critical heat flux (CHF) of water pool boiling, studies on
  curves of, 185
  enhancement, 184–5
  surface coating effect, 185
CVB. *See* Constrained vapor bubble

D

Darcy friction factor, 129
Diffusiophoresis, 151
Direct evaporation condensation (DEC) method, 87, 94
Dispersive equivalent conductivity of medium, 139
Dittus–Boelter correlation, 129–30
Do model of MHP, 17
  See also Steady-state models
Drift velocity model, 120
Dynamic viscosity of nanofluid, 127

E

Effective-medium theory, 112
  for thermal conductivity, 113, 119, 135
Electrohydrodynamic (EHD) heat pipe, 31
  EHD-augmented MHP, 31–2
Electrohydrodynamic spraying system for nanofluids production, 94–5
Energy-intensive optical (LASER) and optoelectronic devices, nanofluid, 187
Epileptic seizures, heat spreader use in, 9, 68
Eulerian approach, 137
Eulerian–Lagrangian model, 139

F

Fluid–particle slip mechanism, 151

G

Gas condensation method, 87
GdS model, 162
Gnielinski correlation, 153
Gold–toluene nanofluid
   enhancement in thermal conductivity with temperature, 111
Gold–water nanofluid
   enhancement in thermal conductivity with temperature, 112
Graetz and Grashof number, 132
Green–Kubo theorem, 124

H

Ha and Peterson perturbation model, 14–15
   See also Steady-state models
Hamilton–Crosser model, 97, 98, 111–12, 162
Heat pipe, 1, 3
   advantages of, 5
   condenser section of, 4
   evaporative section of, 3
   heat spreader, 8
   history of, 1–2
   solar receiver, 10
   and two-phase heat transfer mechanism, 5
   working principle of, 3–4

I

Incompressible suspension momentum equation, 137
Inert gas condensation (IGC) process, 87
Interfacial layer model, 118

K

Kays correlation, 129
k–ε Model, 138
Khrustalev and Faghri numerical model, 12–13
   See also Steady-state models
Kim analytical model, 16
   See also Steady-state models
Kinetic theory-based microconvection and liquid, 123

L

Lagrangian approach, 137
Latent heat of vaporization, 3
Laurate salt, 92–3
Liquid layering theory, 117
Longtin model, 12
   See also Steady-state models

M

Magnus effect, 151
Mass dispersion theory, 141
Mass flux equation and control volume, 146
Maxwell–Garnett model, 117
Maxwell model, 114
MEMS. See Microelectromechanical system
Merit number (Me), 53
MHP. See Microgrooved heat pipe
Microchannel heat sink, 83
Microelectromechanical system, 82
Microelectronic devices and nanofluid, 187
Microgrooved heat pipe, 5
   aim and scope of research on, 2
   applications of, 7
      aircraft temperature control, 10–11
      electronic cooling, 7–8
      fuel cell temperature regulation, 10
      in human diseases remedy, 8–9
      solar energy conversion, 10
      spacecraft thermal control, 9
   designing of, 53–6
   future considerations in research, 64–9
   heat transport limits, 54
      boiling, 63
      capillary (see Capillary limit for MHP)
      condenser, 64
      entrainment, 63
      flooding, 62
      frozen start-up, 64
      sonic, 61
      vapor continuum, 64
      viscous, 63
   MHP catheter, 9
   transient experimental study, 41–3

V-shaped, 6
wire-bonded heat pipe arrays, 9
working principle of, 6–7
See also Heat pipe
Miniature and microgrooved heat pipes, distinction between, 5
Mixture model, 158
Multiwalled carbon nanotubes, 85
MWNTs. *See* Multiwalled carbon nanotubes

N

Nanoelectromechanical system, 82
Nanofluids, 82, 85
  abnormal enhancement of thermal conductivity, 187
  applications, 185–7
  for commercial cooling, 185
  features of, 85–6
  for heat transfer applications, 85
  preparation of, 92–3
    chemical dispersion technique, 93–4
    physical dispersion technique, 93
    sedimentation rate of particles, equation for, 92
    single-step methods, 94–5
Nano heat pipe, 69
Natural convection in nanofluids
  alternate direction implicit (ADI) algorithm, 168
  aspect ratios, 170
  buoyancy-driven heat transfer process, 167
  deterioration in, 165–6
  Dufour effect, 171–3
  electrostatic stabilization mechanisms, 166
  energy equation in, 167
  FIDAP software, solutions for, 168
  Grashof number, 168
  heat and mass flows, 171
  heat transfer behavior, 173
  linear concentration gradient, 171
  liner stability equation, 172
  Richardson number, 173
  separation factor, 172
  Soret effect, 171–3
  stability parameters, 172
  stream function–vorticity formulation, 167, 168
  streamlines and isotherms, 174
  studies
    Nusselt number, 165
    Rayleigh number, 165
  temperature and velocity profiles, 169
  transient and steady heat transfer coefficients, 166
  viscosity model, 173
  weight fraction, 172
NEMS. *See* Nanoelectromechanical system
Notebook computers, heat pipes in, 7

O

ODE. *See* Orientation-dependent etching
Operating and design parameters for MHP
  charging, method of, 46
  effect of surface tension gradient, 47–8
  fabrication, 49–52
  fill charge, 44–5
  friction factor, 48–9
  liquid height and temperature, 47
  optimal fill charge, 45–46
  performance factor, 52–3
  surface roughness, effect of, 48
Orientation-dependent etching, 50
Oxide nanofluids, thermal conductivity measurement, 109

P

Particles
  characterization for nanofluids, 90–2
    dynamic light scattering technique, 90–1
    STM, use of, 91
    TEM, use of, 90, 91
    X-ray diffraction method, 90
  clustering effect on heat transfer augmentation, 116
  drift velocity, 138
  fraction optimization and nanofluid flows, 186
  migration effect in nanofluids, 145
Peclet number
  particle distribution, 150
  temperature and particle size effect on, 149
  viscosity distribution and, 150
  size effect on convective heat transfer, 136
Perkins tube, 2

SUBJECT INDEX

Peterson and Ma model for heat transport in triangular grooves, 13
  See also Steady-state models
Physical dispersion technique for nanofluids, 93
Polymer heat pipes, 9
Polymeric surfactants, 94
Pool boiling
  critical heat flux (CHF) of water, studies on, 184–5
  heat transfer at higher solid particle concentrations, 176–182
    bubble sliding mechanism, 177–8
    curve for, 185
    deterioration in, 182
    dimensionless boiling characteristics, 180
    fouling effect, 182
    heat flux, 181
    Nu–Re_b characteristics, 179
    performance of, 178
    plugging effect, 182
    Rohsenow correlation, 181
    surface characteristics, 177
  heat transfer at lower solid particle concentrations
    enhancement of, 183
    heat flux, 183
    heat transfer characteristics, 183
Prandtl analogy correlation, 153

R

Radial channel
  average heat transfer coefficient, 159
  average wall shear stress, 159
  between heated discs geometries, 155
  local wall temperature and particle volumes, 158
Rayleigh scattering technique
  thermal conductivity of nanofluids, 113
RBC. *See* Rotary blade coupling
Real-time four-wheel drive (4WD) transmission system, 186
Reynolds number and particle concentration, 135, 141
Richardson number, 175
Riffat analytical model, 15
  See also Steady-state models
Rotary blade coupling, 186
Rotor–stator methods, 93

S

Sensitivity analysis
  of steady-state model, 35
  of transient model, 43
Shah correlation, 131
Shear thinning behavior, 126
Sieder–Tate correlation, 132–133
Silica–water nanofluids
  boiling performance of, 183
Silicon microchannel heat sink
  performance using nanofluids as coolants, 186
Single-fluid approach, 154
  equation with dispersion term, 139
Single phase model, 158
Single-walled CNT (SWCNT) nanofluids, 125
Slip mechanisms, 151
Soret and Dufour effects, 140
Soret effect, 120
Space program, heat pipe importance in, 2
Stationary particle model, 121–2
Steady-state experimental studies for MHP, 27
  copper and silver, heat transfer characteristics of, 30
  flat heat pipe, study on, 27
  on maximum applicable power, 30
  maximum heat transport rate, measurement of, 29
  1mm of diameter, study on, 27
  onset and propagation of dry-out point, 28
  thermal optimization with curved triangular grooves, 29
  triangular and rectangular cross sections, 29–30
Steady-state models
  Anand numerical model, 15
  Babin model for trapezoidal MHP, 11
  capillary performance of microgrooves, prediction of, 18
  Cotter model of small MHPs, 11
  1D model of MHP, 15
  Do model of MHP, 17
  1D semianalytical model, 15
  Gerner model, and capillary pressure limit, 12
  Ha and Peterson perturbation model, 13–14
  Khrustalev and Faghri, numerical model of, 12–13

Kim analytical model, 16
Longtin model, 12
numerical thermal model, 16
Peterson and Ma model, for heat transport in triangular grooves, 13
Riffat analytical model, 15
standardized stacked 3D package, development of, 17
Suh and Park model, 17
Suman and Kumar analytical model, 18
Suman numerical model, 18
three-dimensional (3D) model for MHP, 14
Wang model, and porous medium concept, 13
Zhang and Wong, study on MHP by, 16
Stokes–Einstein formula, 122
theory, 92
for viscosity, 138
Stokes law, 139
Stream function–vorticity formulation, 167, 168
  See also Natural convection in nanofluids
Suh and Park model, 17
  See also Steady-state models
Suman and Kumar analytical model, 18
  See also Steady-state models
Surface coating, 94
Synthesis of nanofluids, 87–9
  Brust method, 88, 89, 94
  chemical synthesis method, 88
  by citrate reduction method, 88
  DEC method, 87, 88
  LASER vapor deposition technique, 87
  from oxide particles, 87

T

Temperature effect on nanofluids
  enhancement of thermal conductivity, 111–12
  particle size and conductivity enhancement, 112–13
  temperature oscillation plot, 109
  test cell for temperature oscillation technique, 107–109
  thermal conductivity, measurement of, 109–110
Temperature oscillation technique
  amplitude attenuation, 108
  one-dimensional heat flow and, 107
  periodic boundary conditions, 107–108
  phase shift, 109
  plot, 109
  test cell for, 108
Theorizing concept, 140
  See also Convection in nanofluids
Thermal conductivity
  enhancement in nanofluids, 95
    carbon and polymer nanotube, 104–107
    ceramic, 95–100
    effective-medium theory, predictions of, 113
    metallic nanofluids, 100–104
    test cell for transient hot-wire method, 101–102
  of materials, 83
  of nanofluids
    enhancement, 115
    models, 114
    nanoparticles use, 84–5
    role in cooling technology, 82–3
  of suspensions, 84
Thermal dispersion theory, 130
Thermophoresis, 151
  thermophoretic force, 140
Thermosiphon tube, 2
THWM. *See* Traditional transient hot-wire method
Traditional transient hot-wire method, 95
Transient models study, 33–5
  1D semitransient model, 34
  Sobhan mode, and effective thermal conductivity, 33–4
  transient mathematical model, by Suman and Hoda, 34–5
  Wu and Peterson, numerical model by, 32–3
  See also Microgrooved heat pipe
Turbulent viscosity, 138
Two-phase convection device. *See* Heat pipe
Two-phase liquid cooling technologies, 83
Typical steady-state model for MHP, 18–22
  assumptions in, 19
  cold end, dimensionless boundary conditions at, 22
  corner of section of heat pipe of polygonal shape, 19
  dimensionless parameters, definition in, 21
  dimensionless radius of curvature for heat inputs, 23–4

SUBJECT INDEX

geometrical parameters in, 22
heat flux used for evaporation of liquid, 26–7
hot end, dimensionless boundary conditions at, 22
liquid pressure, estimation of, 20
nondimensionalized equations in, 21
nondimensional parameters in, 20–22
steady mass
  balance for coolant liquid, 20
  balance for vapour, 20
steady-state
  energy balance, equation for, 20
  momentum balance, 20
substrate temperature profiles for heat inputs, 21, 23
variation of
  axial liquid velocity, 24
  axial vapor velocity, 24–5
  liquid pressure, 25, 26
V-shaped pipe, use of, 18
Typical transient models, 35
  dimensionless radius of curvature, profile of, 38–9
  heat flux used by coolant liquid, 35
  liquid pressure and relationship equation, 36
  and nondimensionalization, 36–7
  and temperature profiles of substrate, 38
  unsteady state
    energy balance equation, 36
    mass balance for liquid, 36
    momentum balance, equation for, 36
  and variation of
    axial liquid velocity, 39–41
    liquid pressure, 39, 40

U

Ultrasonic vibration and nanofluid, 188
Uniformly heated tube geometries, 154

V

Vacuum evaporation onto running oil substrate, 87
Vapor-deposited micro heat pipe, 50
VDMHP. *See* Vapor-deposited micro heat pipe
VEROS. *See* Vacuum evaporation onto running oil substrate
Viscosity of CNT-containing nanofluid, 128

W

Wang model and porous medium concept, 13
See also Steady-state models
Wasp model for thermal conductivity, 168
Water–Al_2O_3 nanofluid
  Peclet number and heat transfer coefficient, 133
  shear thinning behavior, 126
Water–CNT nanofluids
  effect of Reynolds number on heat transfer coefficient, 135

X

X-ray diffraction method, 90

Y

Young–Laplace equation, 20, 36, 56